Planning Law Reports 2002

Volume 4

Edited by
Barry Denyer-Green LLM PhD FRICS
of Middle Temple, barrister
and
Navjit Ubhi LLB (Hons)
of Middle Temple, barrister

Yvonne Kirtlan BA (Hons)
Publisher's editor

2002

Routledge
Taylor & Francis Group
LONDON AND NEW YORK

First published 2002 by Estates Gazette

Published 2019 by Routledge
2 Park Square, Milton Park, Abingdon, Oxon OX14 4RN
52 Vanderbilt Avenue, New York, NY 10017

Routledge is an imprint of the Taylor & Francis Group, an informa business

ISBN 13: 978-0-7282-0379-2 (hbk)
ISBN 13: 978-0-7282-0375-4 (set)
ISSN 0954-1977

CONTENTS

CUMULATIVE TABLES - VOLUMES 1-4

CASES REPORTED IN THIS VOLUME

CUMULATIVE TABLES 2002
CASES REPORTED IN VOLUMES 1-4

Numbers refer to volume and page

Buckinghamshire County Council v North West Estates plc (Ch)	**3**-57
Earthline Ltd v Secretary of State for Transport, Local Government and the Regions (CA)	**4**-94
Edinburgh City Council v Scottish Ministers (Ct Sess: OH)	**4**-54
Henry Boot Homes Ltd v Bassetlaw District Council (CA)	**4**-108
Jarmain v Secretary of State for the Environment, Transport and the Regions (No 2) (QB: Admin)	**1**-105
JJ Gallagher Ltd v Secretary of State for Transport, Local Government and the Regions (QB:Admin)	**4**-32
JS Bloor Ltd v Swindon Borough Council (QB: Admin)	**1**-52
Kane v New Forest District Council (CA)	**1**-7
Lidington v Secretary of State for Transport, Local Government and the Regions (QB: Admin)	**3**-37
R (on the application of Adlard) v Secretary of State for the Environment, Transport and the Regions (CA)	**4**-1
R (on the application of Alfred McAlpine Homes Ltd) v Staffordshire County Council (QB: Admin)	**2**-1
R (on the application of Anne) v Test Valley Borough Council (QB: Admin)	**1**-29
R (on the application of Burkett) v Hammersmith and Fulham London Borough Council (HL)	**2**-90
R (on the application of Carlton-Conway) v Harrow London Borough Council (CA)	**3**-77
R (on the application of Chant) v Secretary of State for Transport, Local Government and the Regions (QB: Amin)	**3**-100
R (on the application of Hargrave) v Stroud District Council (CA)	**3**-115
R (on the application of Hossack) v Kettering Borough Council (CA: Cr)	**2**-114
R (on the application of Kides) v South Cambridgeshire District Council (CA)	**4**-66

INDEX OF SUBJECT MATTER

Directives

Waste Framework Directive 75/422/EEC – Planning permission – Waste-disposal sites – S7 of Environmental Protection Act 1990 – Whether breach of duty under s7 – Whether failure to apply objectives in Article 4 of council directive

***R (on the application of Thornby Farms Ltd) v Daventry District Council; R (on the application of Murray) v Derbyshire County Council (CA),* 2-21**

Discontinuance order

Confirmation – S102 of Town and Country Planning Act 1990 – Whether criteria for confirmation akin to confirmation of powers of compulsory purchase – Whether ulterior motive – Whether certain uses wrongly included

***R (on the application of Chant)* v *Secretary of State for Transport, Local Government and the Regions (QB: Admin),* 3-100**

Enforcement control

Enforcement control – Injunctive relief – S187B of Town and Country Planning Act 1990 – Effect of European Convention on Human Rights and Fundamental Freedoms – Art 8 – Breach of planning control – Proper excercise of power under s187B – Whether judges in High Court correctly directing themselves – Whether established approach to s187B compliant with the Human Rights Act 1998

***South Bucks District Council* v *Porter; Chichester District Council* v *Searle; Wrexham County Borough Council* v *Berry; Hertsmere Borough Council* v *Harty (CA),* 3-1**

Enforcement notice

Enforcement – Injunction – S187B of Town and Country Planning Act 1990 – Validity of enforcement notices – Whether validity of enforcement notices in issue where injuntion sought under s187B

***Buckinghamshire County Council* v *North West Estates plc (Ch),* 3-57**

Appeal and application under ss288 and 289 of Town and Country Planning Act 1990 – S289 appeal disposed of – Whether s288 application capable of amendment to s289 appeal – Second s289 appeal hearing – Whether additional issue can be heard at second s289 appeal hearing

***Jarmain* v *Secretary of State for the Environment, Transport and the Regions (No 2) (QB: Admin),* 1-105**

Non-compliance – Prosecution – Change of plea to guilty following judge's ruling – Defence – S179(3) of Town and Country Planning Act

commences" – Conditions stipulating "no development of the site shall begin" until specified steps carried out – Non-compliance with conditions – Works commencing within five-year planning permission validity period – Whether works amounting to valid commencement of development permitted by planning permission – Outline planning permission – Time limits – Whether planning permission implemented within time limits – Whether outline planning permission remaining extant – Whether developer having legitimate expectation that permission treated as implemented – Whether conditions waived

***Henry Boot Homes Ltd v Bassetlaw District Council (CA),* 4-108**

Planning procedures

S64 determination – Town and Country Planning Act 1990 – Waste-treatment plant – Generation of electricity from waste-treatment process – Whether decisions of officer and committee constituting s64 determination

***R (on the application of Reprotech (Pebsham) Ltd) v East Sussex County Council (HL),* 2-60**

Retail development

PPG 6 "Town Centres and Retail Developments" – Sequential approach – Secretary of State disagreeing with inspector's recommendation – Whether Secretary of State correctly applying sequential approach – Whether Secretary of State giving adequate reasons for rejecting inspector's conclusions

***JJ Gallagher Ltd v Secretary of State for Local Government, Transport and the Regions (QB: Admin),* 4-32**

Statutory nuisance

Statutory nuisance – Lime trees – S79 of Environmental Protection Act 1990 – Honeydew excretions causing mould and allergies – Request to serve abatement notice – Whether local authority failing properly to consider request and supporting evidence in deciding to take no action – Whether statutory nuisance existed – Whether "prejudicial to health" – Whether objective or subjective test in determining "prejudicial to health"

***R (on the application of Anne)* v *Test Valley Borough Council (QB: Admin),* 1-29**

Structure plan

Major new housing development policy – Challenge to plan – S31(2) of Town and Country Planning Act 1990 – Whether policy a "general"

CASES JUDICIALLY CONSIDERED

Numbers refer to volume and page

TABLE OF STATUTES

TABLE OF STATUTORY INSTRUMENTS

TABLE OF CIRCULARS

TABLE OF PLANNING POLICY GUIDANCE NOTES

EUROPEAN LEGISLATION

R (on the application of Adlard and others) *v* Secretary of State for the Environment, Transport and the Regions and others

[2002] EWCA Civ 735

COURT OF APPEAL
Simon Brown, Mummery and Dyson LJJ
17 May 2002

Call-in - Planning application - Procedure before local planning authority - Local planning authority resolving to grant planning permission without hearing oral objections - Whether objectors entitled to oral hearing - Secretary of State refusing to call in application - Whether Secretary of State failing to consider omission of oral hearing - Whether Secretary of State erring in failing to call in application - Whether breach of Article 6 of European Convention on Human Rights

In August 1995, the Secretary of State excercised his power, under section 77 of the Town and Country Planning Act 1990, to call in an application for planning permission for a 15,000 capacity football stadium. Following a public inquiry, he granted the permission. That decision was quashed in the House of Lords ([2000] 3 PLR 111) on the ground that no environmental impact statement (EIS) had been provided. In March 2000, the first interested party applied for planning permission for the development of a 30,000 capacity stadium, for which an EIS was provided. Following objections from the appellants, the planning committee carried out a consultation excercise. In February 2001, the committee met and, without allowing oral representations from the objecting appellants, resolved to grant the application. Although requested by the appellants, the Secretary of State refused to call in the application. The appellants' challenge to that decision was dismissed in the court below. In April 2002, the Court of Appeal granted permission to appeal on the grounds: (i) that in deciding not to call in the application, the Secretary of State should have considered whether this was a major development with arguably very important consequences for the local people; and (ii) whether the appellants were entitled as of right to an oral hearing. The issues concerned the rights of the appellants under Article 6 of the European Convention on Human Rights.

Held The appeal was dismissed.

1. The statutory scheme for considering planning applications, as a whole, is plainly compliant with Article 6, and there is no need to resort to the Secretary of State's call-in power to make it so: see [31]. The rights of the objectors to planning applications under Article 6 are not violated if a local planning authority refuse to accord them a

A public inquiry, or any form of oral hearing. This is because a combination of the authority's initial decision-making process and judicial review by the High Court is sufficient to ensure compliance with Article 6: see [46].

2. The nature and scope of the power accorded to the Secretary of State by section 77 of the 1990 Act does not specify any criteria for the exercise
B of the power. The Secretary of State has no duty to consider whether the procedures would be compliant with Article 6 when deciding whether to call in: see [38]. The Secretary of State is given an apparently unfettered discretion to give directions requiring the application for planning permission to be referred to him, instead of being dealt with by local planning authorities, and to give directions to either a particular local
C planning authority or local planning authorities generally. He can exercise the power at any time before the planning authority have "dealt with" the application: see [47].

3. The Secretary of State is not obliged, in deciding how to exercise his call-in powers, to consider departing from his policy to allow parties to put forward their objections at a public inquiry: see [40]. He may have
D regard to this consideration if he thinks it right to do so; it is the type of consideration that the Secretary of State can have regard to or not, as he chooses. The statute permits, but does not require, consideration of the matter *per CREEDNZ Inc* v *Governor-General* [1981] 1 NZLR 172, approved by Lord Scarman in *In re Findlay* [1985] AC 318: see [41].

Cases referred to in the judgments

E *Berkeley* v *Secretary of State for the Environment, Transport and the Regions (No 1)* [2001] 2 AC 603; [2000] 3 WLR 420; [2000] 3 All ER 897; (2001) 81 P&CR 35; [2000] 3 PLR 111; [2001] JPL 58, HL

Bryan v *United Kingdom* A/335-A [1996] 2 EGLR 123; [1996] 28 EG 137; [1996] 1 PLR 47; (1995) 21 EHRR 342

CREEDNZ Inc v *Governor-General* [1981] 1 NZLR 172

F *Findlay, In re; sub nom Findlay* v *Secretary of State for the Home Department* [1985] AC 318; [1984] 3 WLR 1159; [1984] 3 All ER 801, HL

Fredin v *Sweden (No 2) A/283-A* 23 February 1994

Jacobsson v *Sweden (No 2)* (8/1997/792/993) (2001) 32 EHRR 20

R (on the application of Aggregate Industries UK Ltd) v *English Nature; sub nom Aggregate Industries UK Ltd* v *English Nature* [2002] EWHC 908 (Admin); [2003] Env LR 3

G *R (on the application of Alconbury Developments Ltd)* v *Secretary of State for the Environment, Transport and the Regions; R (on the application of Holding & Barnes plc)* v *Secretary of State for the Environment, Transport and the Regions* [2001] UKHL 23; [2001] 2 WLR 1389; [2001] 2 All ER 929; (2001) 3 LGR 38; (2001) 82 P&CR 40; [2001] 2 PLR 76; [2001] JPL 920, HL

R (on the application of Cummins) v *Camden London Borough Council* [2001] EWHC 1116 (Admin)

H *R (on the application of Friends Provident Life Office)* v *Secretary of State for the Environment, Transport and the Regions; sub nom Friends Provident Life Office* v *Secretary of State for the Environment, Transport and the Regions* [2001] EWHC 820 (Admin); [2002] 1 WLR 1450; [2002] JPL 958, AC

R (on the application of Vetterlein) v *Hampshire County Council* [2001] EWHC 560 (Admin); [2002] 1 P&CR 31; [2002] JPL 289, AC

Tower Hamlets London Borough Council v *Begum (Runa)* [2002] EWCA Civ 239; A
[2002] 1 WLR 2491; [2002] 2 All ER 668; [2002] HLR 29, CA

Appeal against a decision of Collins J

This was an appeal by the defendants, Mr William Adlard, Lady Dido Berkeley, Ms Jenny Dearden, Mr Christopher Edwards, Ms Susanna Majendie, Mr Paul Mitchell, Mr John Shannon, Mr Roger Weston and B
Ms Lindsey Carlos-Clarke, against a decision of Collins J, given on 17 January 2002, dismissing their claim and upholding the refusal of the Secretary of State for the Environment, Transport and the Regions to call in an application for planning permission made by Fulham Stadium Ltd, an interested party, pursuant to section 77 of the Town and Country Planning Act 1990. C

Robert McCracken and Gregory Jones (instructed by Richard Buxton, of Cambridge) appeared for the appellants, Mr William Adlard, Lady Dido Berkeley, Ms Jenny Dearden, Mr Christopher Edwards, Ms Susanna Majendie, Mr Paul Mitchell, Mr John Shannon, Mr Roger Weston and Ms Lindsey Carlos-Clarke. D

John Howell QC and James Maurici (instructed by the Treasury Solicitor) represented the respondent, the Secretary of State for the Environment, Transport and the Regions.

Timothy Straker QC and Andrew Tabachnik (instructed by Linklaters & Alliance) represented the first interested party, Fulham Stadium Ltd.

Keith Lindblom QC and Russell Harris (instructed by the solicitor E
to Hammersmith and Fulham London Borough Council) represented the second interested party, Hammersmith and Fulham London Borough Council.

The following judgments were delivered.

SIMON BROWN LJ:

[1] Fulham Football Club's Craven Cottage ground was opened in F
1896. It lies on the north bank of the Thames between Hammersmith and Putney bridges. Its stadium is dilapidated and plainly in need of redevelopment. For some years, its owner, now Fulham Stadium Ltd, the first interested party, has been seeking planning permissions and listed building consents for that purpose. In June 1995, the local planning authority, Hammersmith and Fulham London Borough Council, G
the second interested party, resolved to grant planning permission for a 15,000 capacity, all-seater stadium (the existing capacity of the stadium, not all-seated, being 25,000). In August 1995, however, the Secretary of State exercised his power, under section 77 of the Town and Country Planning Act 1990, to call in the application. Section 77, so far as material, provides: H

> .– (1) The Secretary of State may give directions requiring applications for planning permission… to be referred to him instead of being dealt with by local planning authorities.
>
> (2) A direction under this section –

(a) may be given either to a particular local planning authority or to local planning authorities generally; and

(b) may relate either to a particular application or to applications of a class specified in the direction.

...

(5) Before determining an application referred to him under this section, the Secretary of State shall, if either the applicant or the local planning authority wish, give each of them an opportunity of appearing before, and being heard by, a person appointed by the Secretary of State for the purpose.

[2] In August 1996, following a public inquiry, the Secretary of State granted the planning permission sought. Upon challenge by Lady Dido Berkeley, one of the present appellants, however, that decision was eventually quashed by the House of Lords, on 6 July 2000, on the grounds that there had been no environmental impact assessment and that the Secretary of State had failed to consider properly whether there should have been: see *Berkeley* v *Secretary of State for the Environment, Transport and the Regions (No 1)* [2001] 2 AC 603[1].

[3] Meanwhile, in March 2000, a further application had been submitted. The proposals this time included a 30,000 capacity, all-seater stadium. Unsurprisingly, this has proved highly controversial. The appellants, most of whom live nearby, are among those most strongly opposed. But there are many supporters, too. This time, an environmental impact assessment was produced. A comprehensive consultation exercise was carried out, and a full report was prepared for the planning committee (the committee). On 27 February 2001, the committee met. Some 600 people attended. Oral representations were not permitted. The committee resolved to grant the application.

[4] That decision has never been challenged. Instead, the appellants have sought to persuade (and, by these proceedings, seek to compel) the Secretary of State, as before, to call in the application. One of them, Mrs Jenny Dearden, ended her letter of 28 February 2001 to the Secretary of State:

> I believe the only way to get a reasonable and fair hearing for all parties is for you to call it in to a Public Enquiry. I earnestly ask you to do this.

[5] On 15 March 2001, the appellants' solicitor, Mr Richard Buxton, wrote:

> It seems to us beyond argument that if it was thought right to give the opportunity for objectors to be heard previously, then our client and other local people are entitled to the same opportunity in relation to the new application and proper decision-making process [sic].

---

[1] [2000] 3 PLR 111

[6] On 28 March 2001, the Secretary of State replied, saying that he had A
decided not to call in the application. His letter continued:

> The Secretary of State's policy on call-ins is to be very selective. It is right that in almost all cases the initial decision on whether a development should proceed should be taken by the local planning authority.
>
> Therefore, planning applications are in general only called-in if planning B
> issues of more than local importance are involved, and if those issues need to be decided by the Secretary of State rather than at local level. Each case is considered on its individual merits. However, we have given as examples of cases where the Secretary of State might consider call-in appropriate those which, in his opinion, may conflict with national policies on important matters; could have significant effects beyond their immediate locality; give rise to substantial regional or national controversy; raise significant C
> architectural and urban design issues; or may involve the interests of national security or of foreign Governments.
>
> After very careful consideration, the Secretary of State has concluded, from the information before him that it is right in this case to leave the decision to be taken by the planning authority.

[7] On 6 April 2001, Mr Buxton wrote again, saying that the points that D
he had raised in his letter of 15 March had not been met, and that "it is imperative that the Secretary of State provides full and cogent reasons for his decision". He added:

> One also notes that by failing to call the matter in, the Secretary of State is, in effect, negating the rights of people affected by the development to a hearing, E
> contrary to the Human Rights Act, 1998. A similar point arises in relation to the lack of a hearing in relation to departure from the development plan, which had been subject to inquiry.

[8] On 25 April 2001, the Secretary of State responded, by reference to case law, that he was not required to give reasons. He then repeated his policy very much as set out in his previous letter, stating that he would "in F
general" call in applications only if planning issues of more than local importance were involved, although "each case will be judged on its individual merits". His letter concluded:

> In this instance, after very careful consideration, the Secretary of State has concluded, from all the relevant information before him, including your representations of 15 March, that it was appropriate for the decision as to G
> whether to grant planning consent to be taken by the local planning authority.

[9] It is that decision, the Secretary of State's refusal to call in the application, that it is sought to impugn in these proceedings. The challenge at first instance was advanced on a variety of grounds. Collins J, H
having heard argument for three days in mid-November 2001, comprehensively dismissed the claim in an admirably clear judgment given on 17 January 2002. On 24 April 2002, this court (Laws and Arden LJJ) gave limited permission to appeal on two points. Permission was principally granted to argue that the Secretary of State, in deciding

whether or not to exercise his call-in power, should have considered whether "this being a very major development with arguably very important consequences for the local people... perhaps exceptionally an oral hearing was required", such a hearing not being available to the objectors under the local planning authority's procedures. In addition, however, the court decided not to "shut out" argument that the appellants "were entitled as of right to an oral hearing", although Laws LJ described that as "a very uphill struggle". There being evidence to show that the delay brought about by the challenge had already increased construction costs by some £3m, the appeal hearing was expedited and given a one-day time limit. For my part, I find it surprising that, in those circumstances, the court was presented with bundles containing over 100 authorities.

[10] With that comparatively brief introduction, let me return now to the facts of the case in rather greater detail. These I gratefully take verbatim from the judgment below, passing over the earlier history of the site and starting, in March 2000, with the present planning application.

> 7. In March 2000 the present applications were submitted. They involved the erection of a 30,000 all-seater stadium with ancillary facilities, a new riverside walk and a 7-storey apartment block containing 30 residential units. The existing buildings, including the listed buildings save for the brick façade to the Stevenage Road stand, were to be demolished. In October 2000, the scheme was modified in that the residential block was lowered to 5 storeys and the units reduced to 16 and there was some reduction to the proposed length of the building. There were some modifications to the design of the riverside walk. The application had been accompanied by a full environmental statement which was supplemented in October 2000. That statement was submitted for review and grading to the Institute of Environmental Management and Assessment and overall it was considered to be a good statement, containing only minor omissions and inadequacies (which were identified), and in some respects was excellent. In addition, the LBH gave the proposals extensive publicity and carried out comprehensive consultation. That exercise was also carried out in relation to the similar proposal lodged in July 1999. This was overtaken by the March 2000 applications largely because the LBH in September 1999 required the applicants to produce an Environmental Impact Assessment which was to and did include the following:–
>
> "(i) The impact of the physical scale of the development in relation to surrounding land uses and occupiers.
>
> (ii) The impact of the use of development on match days, including how other uses within the site operate during that time, how people would arrive at and depart from the site and the impact and extent this would have on the surrounding area including adjoining Boroughs, in terms of noise, disturbance, litter and transport (including issues such as traffic generation, on and off street parking, existing/potential public transport provision, pedestrian movements etc). Also any additional impact connected with the emergency evacuation of the stadium.
>
> (iii) The impact of the use of the development on non-match days, the type, frequency and hours of use of all the proposed activities and the numbers likely to be involved and the impact of this on the surrounding area in terms of noise, disturbance, litter and transport.

(iv) The impact of the development on the historical and architectural interest of the remaining listed buildings, the character and appearance of the conservation areas and the justification for the demolition of the listed buildings. The impact of the development on Bishops Park as a landscape of historical/cultural/archaeological interest.

(v) The impact of the development on the ecology of Bishops and Stevenage Parks.

(vi) The impact of the development on the archaeology of the application site which includes the foreshore of the River Thames.

(vii) The visual impact on the development on the River Thames from the river, from riverwalks on both banks, from Hammersmith and Putney Bridges and from other points across the river.

(viii) The impact on the ecology of the River Thames.

(ix) The impact of the development on the hydrology and flood defences of the River Thames.

(x) The impact of the development on the recreational, leisure and commercial uses of the River Thames."

8. The responses to the consultation exercises are summarised in the very full and detailed report which was prepared for the Committee which considered the applications on 27 February 2001. It runs to 115 pages together with a further 191 pages of annexes. Annex 2 (145 pages) contained the consultation responses which were summarised in the body of the report in Annex 1. It is not nor could it be suggested that the report failed fairly to identify the objections which had been put forward or to set out all relevant considerations which the committee must take into account, including the fact that the proposals were not in accordance with the UDP in the respects to which I have already referred [I should interpolate that the Unitary Development Plan, adopted in December 1994, included as Site Policy 19:

"(a) Retention and enhancement of listed buildings in association with football or other spectator sport or entertainment as appropriate in the location;

(b) Provision of a public Riverside walk with links inland as appropriate;

(c) Residential development reasonably necessary to enable the achievement of these objectives and in accordance with normal housing and environment policies."

The report recommended that the applications should be approved and planning permission granted subject to a satisfactory s106 agreement and to a number of conditions to ensure that there should be as much compliance as possible with the relevant policies in the UDP and that the adverse impact of the development on local residents should be kept as small as possible. In particular, it was proposed (and the Committee agreed) to include the following conditions:

"7. The stadium's primary purpose shall be for the playing and watching of Association Football and the playing area and terracing and seating overlooking the playing area shall only be used for the following purposes:

(a) The home matches of only one professional club for the playing of Association Football.

(b) Events within Use Class D2(e) for children and educational establishments.

To ensure that the development does not result in conditions prejudicial to the amenities of the surrounding area from additional traffic, noise and

A disturbance which may be generated by the development contrary to UDP policies, EN21, TN14 and TN15 as a consequence of more frequent crowd-generating or noise than one club's football matches.

...

9. On non-match days the total number of patrons using all the facilities of the stadium shall not exceed 150 at any one time unless it is for an event under Use Class D2 (e) for children and educational establishments."

B There were other conditions (such as a prohibition use of the stadium between 11pm and 9am and that the restaurant must only be used as such and the café must not be used after 8pm except on match days) which were designed to keep disturbance to a minimum, including restrictions applicable to the demolition and construction involved.

9. It is to be noted that there were a large number of responses which C supported the applications. Overall in terms of numbers more responses in support were received than in opposition. However, many letters in support were what are described as standardised. But it is clear that there is considerable support for the proposals as well as objection to them. Furthermore, it is important to bear in mind that the development will not bring a new football ground to a residential area where none existed before. Those who live nearby either came to or have spent all their D lives close to a football ground which, depending on the future of the club, has attracted a more or less substantial number of fans. The problems of parking and congestion on match days have always existed. In addition, there have been recent concerns, reflected in correspondence, about the loutish behaviour of fans which has intimidated local residents and which the police have not seemed either willing or able to control. One of the matters to be covered by the s106 agreement involves the installation of E CCTV in the streets immediately adjoining the stadium. It is convenient to set out the brief summary of the matters raised against and in support of the proposals. These are in Paragraphs 4.13 to 4.15 of the officer's report, which read:

"4.13. The main representations put forward by objectors are:

(i) The demolition of the listed buildings, involving the loss of Craven F Cottage in its entirety and the demolition of all but the façade of the Stevenage Road stand is unacceptable and unjustified.

(ii) The scale and design of the Stadium is unacceptable in relation to the site and the surrounding area.

(iii) The proposals would fail either to preserve or enhance the character and appearance of the conservation areas within which the site lies and adjoins. The site lies within the Fulham Reach Conservation Area and adjoins G the Bishops Park and Crabtree Conservation areas.

(iv) The height, massing and visual appearance of the proposal, because of the site's prominent position on the River Thames, would have an unacceptable impact.

(v) The development would be detrimental to residential amenity by reason of its visual impact, traffic generation and associated effects upon air quality.

H (vi) The proposed development is contrary to UDP policies, particularly Site Policy 19.

(vii) The highways impact on the proposals would be unacceptable. The use of the existing stadium puts pressure on the highway network and upon parking. The redevelopment proposals have the potential to more than double the attendance. The impact of the additional cars would be felt in

Fulham and outside the Borough. There would be 26,000 HGV movements during the construction period which would adversely impact on residential amenity and the highway network. A

(viii) The public transport system would be incapable of adequately catering for the increase in passenger movements.

(ix) The development would severely prejudice the function of the River Thames as an ecological resource and as a recreation facility.

(x) The stadium might become a 'white elephant' if the football club either fail to achieve promotion to the Premier League or fail to maintain Premier League status if promotion is attained. B

(xi) The applicants have failed to consider all alternatives to the proposals, including the feasibility of the football group being redeveloped as a phased programme which would have different environment impacts.

(xii) The need case for a new stadium of 30,000 capacity has not been proven nor has the need for the housing block. C

(xiii) There would be unacceptable encroachment onto the River Thames which is public open space and should be protected from development. Any development should be contained within the site.

4.14 Additionally objectors criticised the Environment statement. They contend that it is deficient in its assessment of a number of key issues including highways and transportation, impact on the river and consideration of alternatives. D

4.15 The main points made by supporters of the development are as follows:

(i) The proposal would be an improvement visually over what is on the site at present.

(ii) The stadium has to be brought in line with safety requirements and football guidelines and this is an opportunity to provide a well designed and up to date new stadium. E

(iii) It is important to retain the club on the site as it is part of the Borough's heritage.

(iv) The development would introduce a valuable new section of river walk and enhanced river views.

(v) The football club makes a significant contribution to the local community and the development would enable this contribution to continue and to be enhanced by securing Fulham Football Club's future at its historic home. F

(vi) The development would bring employment to the area."

10. The Fulham Alliance had submitted a fully reasoned objection to the proposals through expert advisers. There was a further reasoned objection from Thamesbank, which dealt with the social and environmental damage which would be caused by the proposals, particularly to the River Thames. The Fulham Alliance document made the point that it was "not convinced that the football club have fully explored the possibility of pursuing a smaller all seater stadium, for example accommodating 20,000-25,000 spectators". The present capacity of the ground is 25,000, but that is because it is not all seated and it will not be approved by the Football Association unless it is. That follows the recommendations made in the Taylor Report on the Hillsborough disaster. Fulham Stadium Ltd assert that a ground with a capacity of 30,000 is the minimum which will be financially viable in the long term. In the report, the officers have said that they are not convinced of this or that a development with a smaller capacity ground would necessarily mean that the Club would not have a future. But they say in Paragraph 15.22: G H

A "What is apparent is that, if the proposed stadium were built, Fulham Football Club would be likely and probably more likely to remain on its historic site into the foreseeable future. Additionally, it is likely that the existence of the new stadium would contribute towards future sporting success though, of course, it would in no way guarantee it."

11. The meeting of the Committee on 27 February 2001 attracted a full house: some 600 people apparently attended it. There were complaints made B that the Committee seemed to have made their minds up in advance and that objectors were not able to make any representations. English Heritage, whose views on the listed buildings consent were of great importance, had stated that their support for the proposal:

"has been and continues to be made on the basis that the Council shall have properly satisfied itself that the size and capacity of the proposed stadium is C the minimum to ensure the viability of the Football Club development and the provision of the benefits referred to in the supporting documentation."

At the meeting, the Assistant Director (Development Services) told the Committee that the officers had concluded that, because of the intangibles surrounding the financing of football clubs and a scheme such as the one proposed, it was most unlikely that any conclusive evidence could ever be forthcoming. Officers, he said, therefore took the view that notwithstanding D the qualification contained in the English Heritage letter, the visual, historic and aesthetic qualities of the proposals remained unchanged and the recommendations to approve the applications should stand. There can be no doubt that English Heritage's concerns were properly put before the Committee which was, in the exercise of its judgment, entitled to decide that the scheme could nevertheless be approved.

E [11] Those, then, are the facts concerning the development itself and the circumstances in which it came to be approved by LBH's resolution of 27 February 2001. I have already sufficiently indicated the circumstances in which the Secretary of State thereafter declined to exercise his call-in power. I can turn now therefore to the grounds of appeal for which leave was given, these grounds raising, as it seems to me, the following issues:

F (i) Given that the Secretary of State has declined to call in this planning application, have the appellants' Article 6 rights been breached by LBH's failure to accord them any form of oral hearing, let alone a public inquiry?

(ii) If so, was the Secretary of State therefore obliged to call in the application so as to prevent the breach?

(iii) Even if the appellants had no right to an oral hearing of their G objections, ought the Secretary of State nevertheless to have considered whether exceptionally to call in the application so as to accord them such a hearing?

(iv) If so, did he consider this?

[12] It will readily be seen that issues (i) and (ii) go to the subsidiary ground of appeal for which permission was given only in the slipstream H of the other ground. They are, however, issues of the utmost importance to the entire domestic planning process, and, logically, they fall to be considered first. It will also be appreciated that for the appellants to succeed on this first and most fundamental ground of appeal, they must succeed on issue (ii) as well as issue (i). Given that, for reasons to which I shall come, issue (ii), to my mind, plainly falls to be resolved

against the appellants, it is tempting not even to address the rather more complex first issue. There is, after all, no separate challenge to LBH's resolution to grant planning permission, surprising though that may seem. Having, however, heard full argument on issue (i), and given its obvious wider importance and the (at least theoretical) possibility that the appellants might hereafter, were the present challenge to fail, seek to challenge LBH's proposed formal grant of planning permission, I think it right to decide the point. A B

Issue (i)

[13] So far as presently material, Article 6(1) provides:

> In the determination of his civil rights and obligations... everyone is entitled to a fair and public hearing... by an independent and impartial tribunal established by law. C

[14] The respondents and interested parties, while reserving their position in the event of any further appeal, have been prepared to assume, for the purposes of the present appeal, that the grant of planning permission here will be directly decisive of the civil rights of one or more of the appellants (albeit less obviously in their case as third-party objectors than, say, in the case of site owners) so that the decision as to whether or not to grant planning permission engages their rights under Article 6. D

[15] Issue (i) can, therefore, be restated as follows: Is the appellants' entitlement to "a fair and public hearing" by "an independent and impartial tribunal" satisfied by the English planning system, which, in the case of objectors, allows the local planning authority to grant planning permission without having afforded them any opportunity of an oral hearing, subject only and always to this court's supervisory jurisdiction on a judicial review application? E

[16] Perhaps somewhat surprisingly, the local planning authority themselves are not to be regarded as an independent and impartial tribunal. Independent, for this purpose, means independent of the executive as well as of the parties. So much is accepted. The question as to whether or not Article 6 is satisfied, however, falls to be considered by reference not merely to the initial decision-making process but also in the light of the High Court's review jurisdiction. The proper approach was recently clarified by Laws LJ's valuable judgment (with which Lord Woolf CJ and Dyson LJ agreed) in *Tower Hamlets London Borough Council v Begum (Runa)* [2002] EWCA Civ 239[1]. The court there was concerned with whether the procedure for a local housing authority's internal review of their decisions on homelessness, such reviews being subject only to appeal in point of law to the county court, complies with Article 6. Laws LJ said: F G H

[1] [2002] 1 WLR 2491

> **40.** As I have shown, the extent to which the first instance process may be relied on to produce fair and reasonable decisions is plainly an important element. But it is not to be viewed in isolation. The matter can only be judged by an examination of the statutory scheme as a whole; that is the necessary setting for any intelligent view as to what is fair and reasonable. Where the scheme's subject matter generally or systematically involves the resolution of primary fact, the court will incline to look for procedures akin to our conventional mechanisms for finding fact: rights of cross-examination, access to documents, a strictly independent decision-maker. To the extent that procedures of that kind are not given by the first instance process, the court will look to see how far they are given by the appeal or review; and the judicial review jurisdiction (or its equivalent in the shape of a statutory appeal on law) may not suffice. Where however the subject matter of the scheme generally or systematically requires the application of judgment or the exercise of discretion, especially if it involves weighing of policy issues and regard being had to the interests of others who are not before the decision-maker, then for the purposes of Article 6 the court will incline to be satisfied with a form of inquisition at first instance in which the decision-maker is more of an expert than a judge (I use the term loosely), and the second instance appeal is in the nature of a judicial review. It is inevitable that across the legislative board there will lie instances between these paradigms, sharing in different degrees the characteristics of each. In judging a particular scheme the court, without compromise of its duty to vindicate the Convention rights, will pay a degree of respect on democratic grounds to Parliament as the scheme's author.
>
> ...
>
> **43.** I should indicate moreover that although there were sharp issues of primary fact falling for determination in the present case, that is not a necessary feature in a section 202 review, and certainly not a systematic one. As often as not there will be no real question of fact, and the decision will turn on the weight to be given to this or that factor against an undisputed background... Now, clearly the statutory scheme is either compliant with Article 6 or it is not. Its compliance or otherwise cannot vary case by case, according to the degree of factual dispute arising. That would involve a wholly unsustainable departure from the principle of legal certainty. In my opinion, judged as a whole, this statutory scheme lies towards that end of the spectrum where judgment and discretion, rather than fact-finding, play the predominant part.

[17] What, then, of the planning process? Where in the spectrum does this statutory scheme lie? To my mind, there can be only one answer to that question, and it is the same answer as *Begum* gave with regard to the homelessness legislation, namely "towards that end of the spectrum where judgment and discretion, rather than fact-finding, play the predominant part". Accordingly (see para 40 of *Begum*), "the court will incline to be satisfied with a form of inquisition at first instance at which the decision-maker is more of an expert than a judge... and the second instance appeal is in the nature of a judicial review".

[18] It is against this background that I now come to consider Mr Robert McCracken's argument that it is none the less imperative, for

compliance with Article 6, that some form of oral hearing, if not indeed a full public inquiry, be accorded by the decision-making body in every case. The argument, I should note, relates most specifically, perhaps exclusively, to those who object to the grant of planning permission. If planning permission is refused, the applicant has a right of appeal to the Secretary of State that will secure him, if he wishes, a public inquiry. That is the position too, of course, with regard to enforcement proceedings. It is objectors only who have no right to any form of oral hearing before the local planning authority, and no right of appeal. Is this fatal to the scheme of the legislation?

[19] In submitting that it is, Mr McCracken places heavy reliance upon *R (on the application of Alconbury Developments Ltd)* v *Secretary of State for the Environment, Transport and the Regions* [2001] UKHL 23; [2001] 2 WLR 1389[1], and, in particular, those of the speeches that underline the importance of public inquiries. Among the passages relied upon are these by Lord Slynn at para 46:

> The fact that an inquiry by an inspector is ordered is important. This gives the applicant and objectors the chance to put forward their views, to call and cross-examine witnesses. The inspector as an experienced professional makes a report, in which he finds the facts and upon which he makes his recommendations.
>
> ...

and Lord Hoffmann at para 117:

> It is only when one comes to findings of fact, or the evaluation of fact, such as arise on the question of whether there has been a breach of planning control, that the safeguards are essential for the acceptance of a limited review of fact by the appellate tribunal."

The safeguards there referred to are those described by the ECtHR in *Bryan* v *United Kingdom* A/335-A (1995) 21 EHRR 342[2], which include that of a public inquiry before an inspector. Lord Clyde said, at para 152:

> In the civil context the whole process must be considered to see if the article [Article 6] has been breached. Not every stage need comply. If a global view is adopted one may then take into account not only the eventual opportunity for appeal or review to a court of law, but also the earlier processes and in particular the process of public inquiry at which essentially the facts can be explored in a quasi-judicial procedure and a determination on factual matters achieved.

At para 157, he said:

> So far as the manner in which the decisions will be taken is concerned it is to be noticed that in each case there will be a public inquiry before an

[1] [2001] 2 PLR 76
[2] [1966] 1 PLR 47

> inspector. That will be an occasion for the exploration of the facts, including the need and desirability of the development.

[20] Mr McCracken's reliance upon these passages is misplaced. Quite apart from the fact that his argument plainly proves too much – is it seriously to be suggested that anyone objecting to a planning application has the right to a public inquiry before an inspector? – it entirely overlooks the context in which those views were expressed. The focus in *Alconbury* was on the requirement under Article 6 for "an independent and impartial tribunal", and, in particular, the propriety of the Secretary of State himself taking the planning decisions in question: in two of the cases having himself called in the applications (the very opposite of the position on the present appeal) and in the third case being seised of a compulsory purchase order in respect of land sought by his own department for a road improvement scheme. The cases were in no way concerned, as this one is, with the requirements of "a fair and public hearing". The speeches simply never addressed the question of whether any form of oral hearing was required. As to the importance generally of the "safeguards" identified in *Bryan*, it is necessary to read the first part of para 117 of Lord Hoffmann's speech:

> If, therefore, the question is one of policy or expediency, the "safeguards" are irrelevant. No one expects the inspector to be independent or impartial in applying the Secretary of State's policy and this was the reason why the court said that he was not for all purposes an independent or impartial tribunal. In this respect his position is no different from that of the Secretary of State himself. The reason why judicial review is sufficient in both cases to satisfy Article 6 has nothing to do with the "safeguards" but depends upon the *Zumtobel* principle of respect for the decision of an administrative authority on questions of expediency.

[21] Lord Hoffmann then went on to make the distinction between that type of administrative decision (typically, the grant of planning permission, where the decision turns on "questions of expediency") and enforcement notice appeals, which typically turn on questions of fact: a distinction that Lord Hoffmann had earlier made in paras 90 and 110.

[22] In short, *Alconbury*, to my mind, says nothing as to whether objectors are entitled to any form of oral hearing. What, then, of other domestic authorities? We were shown a series of first instance decisions, none of which supported Mr McCracken's argument, and several of which tend strongly against it. I need touch on them only briefly, exceedingly long although most of them are.

[23] Sullivan J in *R (on the application of Vetterlein)* v *Hampshire County Council*[1] said, at para 68:

> The special meeting was held in public. The agenda was available to members and to the public beforehand. In deciding whether there has

[1] [2002] EWHC 560 (Admin); (2002) 1 P&CR 31

> been a breach of Article 6(1) the procedures have to be looked at in their A
> entirety, including the earlier opportunities to make representations during the consultation process and the subsequent right to seek relief by way of judicial review if the Council errs in law. A "fair" hearing does not necessarily require an oral hearing, much less does it require that there should be an opportunity to cross-examine. Whether a particular procedure is "fair" will depend upon all the circumstances, including the nature of the
> claimant's interest, the seriousness of the matter for him and the nature of B
> any matters in dispute.

[24] True it is, as Mr McCracken points out, that Sullivan J had already by then concluded that the claimants' interest was so remote that no Article 6 rights were engaged, but, in para 71 he expressed himself
"satisfied that even if Article 6(1) did entitle the claimants to a fair and C
public hearing, the procedures adopted by the county council, when looked at in totality, did afford them just that opportunity.

[25] Forbes J's decisions in *R (on the application of Friends Provident Life Office)* v *Secretary of State for the Environment, Transport and the Regions* [2001] EWHC 820 (Admin)[1] and *R (on the application of Aggregate*
Industries UK Ltd) v *English Nature* [2002] EWHC 908 (Admin)[2] and D
Ouseley J's decision in *R (on the application of Cummins)* v *Camden London Borough Council* [2001] EWHC 1116 (Admin) all adopt that same approach. *Friends Provident* and *Cummins* were both concerned, like the present case, with objectors' challenges to the Secretary of State's refusal to call in planning applications. Both considered, as I am considering, the
sufficiency, for Article 6 purposes, of the local planning authority's E
procedures when taken together with the court's supervisory jurisdiction. Ouseley J, in *Cummins*, concluded, in para 341 (of a 374-paragraph judgment refusing applications for permission to move for judicial review):

> Decisions as to whether or not to grant planning permission are policy or F
> "expediency" decisions which call for a range of planning judgments. These are decisions entrusted to democratically accountable bodies, because of the policy content. Article 6(1) is complied with because of the availability of judicial review. On that review, the Court has the full jurisdiction to deal with procedural and substantive errors of law. Its jurisdiction therefore permits a
> challenge to be brought on the basis that a material fact has been ignored or G
> misunderstand or that a view of the facts or their evaluation has been reached unreasonably. … I do not consider that the comments in the speeches in *Alconbury* and of Lord Hoffmann in particular can be taken to suggest that findings of fact and the evaluation of fact involved in the decision whether or not to grant planning permission, require the interposition of an inspector in order for the limited review of fact by the court to satisfy
> Article 6. H

[1] [2002] 1 WLR 1450

[2] [2003] Env LR 3

[26] Forbes J in both *Friends Provident* and *Aggregate Industries* (giving express consideration in the latter to the Court of Appeal's judgment in *Begum*) reached a similar conclusion.

[27] True it is that, in all these cases, the objectors were given some limited opportunity to make oral representations. Paragraph 86 of Forbes J's judgment in *Friends Provident* notes that the objectors' advisors made submissions, both in writing and orally, at the planning committee's meeting. Ouseley J in *Cummins* records, in paras 130 and 143, that oral deputations by objectors were heard initially by the relevant subcommittee and then again by the full council. Forbes J in *Aggregate Industries* – a challenge to the confirmation of certain land as a site of special scientific interest – records in para 103(vi) that the applicant's solicitor made oral representations at the relevant council meeting and was "given a fair and public hearing on that occasion". Nothing in the judgments, however, suggests that those brief oral hearings were central to the court's conclusion that the planning system as a whole satisfies the requirement under Article 6 for "a fair and public hearing" – or, indeed, that, but for those hearings, the court would have struck down the individual decisions as unfair.

[28] I pass, therefore, to the final authorities upon which Mr McCracken seeks to rely with regard to issue (i): two judgments of the ECtHR in *Fredin* v *Sweden (No 2) A/283-A* 23 February 1994 and *Jacobsson* v *Sweden (No 2)* (8/1997/792/993)[1], which I can deal with really very briefly. In paras 21 and 22 of its judgment in *Fredin*, the court stated that, under its existing case law, it was clearly established that "in proceedings before a court of first and only instance", the right to "a public hearing" entailed an entitlement to an "oral hearing". The critical question arising there was whether the fact that the applicant was denied an opportunity to present oral argument before the Supreme Administrative Court, the first and only tribunal seised of the applicant's challenge to the underlying administrative decision, breached Article 6(1). The court held that it did: the applicant had been guaranteed a right to an oral hearing and had had none.

[29] That decision was applied in *Jacobsson*, although, in the event, the court there reached a different conclusion:

> 49. …the Court does not find on the evidence before it that the applicant's submissions to the Supreme Administrative Court were capable of raising any issues of fact or of law pertaining to his building rights which were of such a nature as to require an oral hearing for their disposition… on the contrary, given the limited nature of the issues to be determined by it, the Supreme Administrative Court, although it acted as the first and only judicial instance in the case, was dispensed from its normal obligation under Article 6(1) to hold an oral hearing.

---

1 (2001) 32 EHRR 20

[30] Those decisions cannot, in my judgment, avail the appellants: it A
hardly needs pointing out that had they challenged LBH's resolution by
judicial review, they would plainly have been entitled to an oral hearing of
their application. Neither *Fredin* nor *Jacobsson* say anything to suggest that
Article 6(1) confers the right to be heard orally by the administrative
decision-making body.

[31] For my part, I can find no warrant, whether in domestic or in B
Strasbourg jurisprudence, for concluding that where, as in *Begum* and as
again here, the administrative decisions taken at first instance are
generally likely to turn on questions of judgment and discretion, rather
than on findings of fact, the statutory scheme must provide for an oral
hearing at that initial stage. On the contrary, I have reached the clearest
conclusion that the statutory scheme as a whole is plainly compliant with C
Article 6, and that there is no need to resort to the Secretary of State's
call-in power to make it so.

[32] With regard to issue (i), I add just this. The remedy of judicial
review, in my judgment, amply enables the court to correct any injustice
it perceives in an individual case. If, in short, the court were satisfied
that exceptionally, on the facts of a particular case, the local planning D
authority had acted unfairly or unreasonably in denying an objector
any, or any sufficient, oral hearing, the court would quash the decision
and require such a hearing to be given. This presents no difficulties:
Mr McCracken disputes neither the authority's power to conduct
such a hearing nor the court's power to order it. I should make it plain,
however, that I am by no means persuaded that any oral hearing was E
required on the facts of the present case. Quintessentially, the decision
whether or not to permit this development (and the departure from the
development plan that it represents) involves questions of discretion and
planning judgment, rather than the resolution of primary fact.
When invited to identify the most acute factual issues in the case,
Mr McCracken could find nothing more telling than: (a) whether the F
development would result in 3,744 more parked cars (the appellants'
estimate) or a maximum of 2,500 parked cars (the respondent's rival
case); and (b) whether only 1,080 cars could be accommodated within
the suggested area (as the appellants contend) or as many as 1,800 to 2,300
(the respondent's estimate).

G

Issue (ii)

[33] Having regard to my conclusion upon the first issue, it could
perhaps be said that issue (ii) now becomes moot. Recognising, however,
that it is the Secretary of State's refusal to call in this application that is
directly under challenge here, I shall address it none the less.

[34] Upon what basis do the appellants seek to impose upon the H
Secretary of State a duty to exercise his section 77 power so as to prevent
a violation of Article 6? Mr McCracken argues that the call-in power is
"part of the composite process which taken as whole achieves compliance
with Article 6". To this end, he relies upon a single sentence from Lord
Clyde's speech in *Alconbury*:

A

> 159 ...Parliament, democratically elected, has entrusted the making of planning decisions to local authorities and to the Secretary of State with a general power of supervision and control in the latter.

[35] Founding himself upon that broad *dictum*, Mr McCracken submits that the Secretary of State has a supervisory role comparable to that of the court on judicial review but to be exercised before the review jurisdiction

B

is invoked, judicial review being a remedy of last resort. The Secretary of State, he submits, is the "primary decision-maker as to whether there should be a public hearing".

[36] I have no hesitation in rejecting this argument. There are, I think, two central points to be made with regard to it. First, the Secretary of State's obligation under section 6 of the Human Rights Act 1998 is not

C

himself to act incompatibly with a Convention right; he is not obliged to ensure that other public authorities themselves act compatibly. Second, it is one thing to accept, as the Secretary of State does, that the decision whether or not to grant planning permission in this case directly affects the appellants' civil rights so as to engage Article 6; it is quite another to suggest that Article 6 is engaged by the exercise of the section 77 discretion

D

whether or not to call in the application. The latter is simply a procedural decision as to which of two public authorities should take the substantive decision. In short, it seems to me perfectly clear that Article 6 rights are not engaged by the exercise of the Secretary of State's section 77 discretion. Similarly, even assuming that the appellants' Article 8 rights are engaged with regard to the substantive planning decision, there can be no question

E

of their being interfered with by the exercise of the Secretary of State's call-in power.

[37] It is to be noted that Ouseley J rejected precisely the same argument in *Cummins*:

> 360. [Claimants' counsel] recognised the force of [respondent counsel's]

F

> next submission that even if [the claimant's] civil rights were engaged and the way in which the Council dealt with the applications involved a breach of Article 6, it was not the duty of the Secretary of State to judge the adequacy of the procedures adopted by the Council and then to intervene to ensure that the procedures which it adopted complied with Article 6. The Secretary of State has no power to do that. It is for the Courts to review the decisions of public bodies and not the Secretary of

G

> State. It is for the Court which tells the Council what it must do as a public body to comply with the obligations in section 6(1) of the Human Rights Act 1998 so as to avoid acting in ways which are in breach of a convention right. If the Courts conclude that there has been a breach of Article 6 by the Council it will tell the Council what it has to do in order to comply with Article 6 and it would then be for the Council to comply with those directions. I accept [respondent counsel's] submission that the

H

> combination of the Court's direction as to the procedure to be followed and the obligation of the Council to comply with that direction provide a complete system by which compliance with Article 6 is then ensured. None of that procedure involves the Secretary of State's intervention. It follows that the Secretary of State does not act incompatibly with Article 6 rights if he declines, as he did in this case, to call the case in for his decision. There is no

> duty under which he himself is obliged, by virtue of section 6(1) of the Human Rights Act to call in the planning applications.
>
> 361. The Secretary of State has no duty to substitute his supervision of the Council's procedures for that of the High Court. It is the High Court, in the exercise of its power of review, which has the role of authoritatively deciding what minimum standards of fair procedure are required by Article 6 as a matter of law in the particular case. This argument was accepted was Forbes J in the *Friends Provident* case.

[38] It follows from all this that, in my judgment, Collins J was correct to conclude as he did:

> 28... the Secretary of State has no duty to consider whether the procedures would be compliant with Article 6 in deciding whether or not to call in. It is matter for the court to decide whether overall the procedures were compliant and fair. I am satisfied in this case that they were. But the challenge in my view should have been made to the LBH's decision, not the Secretary of State's.

Issue (iii)

[39] I have already indicated my view that the Secretary of State was entitled to exercise his section 77 discretion without regard to whether or not, were he to decline to call in the application, the High Court, on judicial review, would view the objectors as having been unfairly treated and so itself order an oral hearing. Ought he nevertheless to have considered whether or not the desirability of there being an oral hearing was such that, exceptionally, he should call in the application, even though under his policy he would not otherwise do so? Was he legally obliged to consider departing from his policy so as to ensure that the objectors became entitled to a hearing, such being, in practice, the invariable consequence of call-in? Was this, in short, a material consideration of the kind that the Secretary of State was bound to take into account when exercising his discretion?

[40] For my part, I think not. As Lord Slynn recorded in *Alconbury*, the Secretary of State in fact calls in only around 130 of some 500,000 planning applications made annually. To alter, or depart from, his present policy (described in [6]) to accommodate cases where objectors have procedural grievances would significantly distort the approach hitherto taken to the exercise of this power. Mr McCracken makes the point that the present application, involving, as it does, a substantial departure from the development plan, was one that LBH were obliged to notify to the Secretary of State, pursuant to para 3(c) of the Town and Country Planning (Development Plans and Consultation)(Departures) Directions 1999, for him to consider whether to exercise his call-in power. He furthermore points out that anyone objecting to policies and proposals in the draft development plan would, at that stage, have been entitled to be heard at a public inquiry. I nevertheless remain unpersuaded that, even in this relatively limited category of cases, the Secretary of State is bound to consider departing from his policy simply so as to enable third parties to put their objections at a public inquiry.

[41] That is not to say, however, that the Secretary of State cannot have regard to this consideration if he thinks it right to do so. Rather, in my judgment, it is a consideration of the kind that the Secretary of State can have regard to or not, as he chooses. The statute permits, but does not require, consideration of the matter: see *CREEDNZ Inc* v *Governor-General* [1981] 1 NZLR 172, approved by Lord Scarman in the House of Lords in *In re Findlay* [1985] AC 318, at pp333-334.

Issue (iv)

[42] Did the Secretary of State in any event consider whether exceptionally to call in the application, so as to accord the objectors an oral hearing? Mr McCracken submits not, urging that it would be wrong to conclude that he did since he refused to give any proper reasons for declining to call in the application. I recognise that certain of the phrases used in the Secretary of State's letters of 28 March and 25 April 2001 – "each case is considered on its individual merits", "after very careful consideration", and such like – appear somewhat standardised (appear, indeed, to have been used in similar letters sent in the earlier cases that establish that reasons – beyond those implicit in saying that the case does not fall within the Secretary of State's policy for calling in applications – are not required for such decisions). There nevertheless seems to me no sufficient reason here for declining to take at face value the Secretary of State's assertion, at the end of his letter of 25 April 2001, that he had given careful consideration to "all the relevant information before him, including [Mr Buxton's] representations of 15 March", representations that expressly contended that the objectors were entitled to "a hearing in relation to departure from the development plan, which had been subject to inquiry".

[43] Even, therefore, were I to conclude that the Secretary of State was bound to consider calling in the application, specifically so as to accord the appellants an oral hearing of their objections, I would hold, on the evidence before the court, that in fact he did so.

[44] It follows that, in my judgment, each of the four issues raised before us falls to be determined against the appellants. Whatever sympathy one may feel for them in the face of this impending development, they have failed to establish any basis for a legal challenge. I would dismiss this appeal.

MUMMERY LJ:

[45] For the reasons set out by my lords, Simon Brown and Dyson LJJ, I, too, would dismiss this appeal.

DYSON LJ:

[46] I agree that this appeal should be dismissed, and for the reasons given by Simon Brown LJ. As he explains in his analysis of issue (i), the rights of objectors to planning applications, under Article 6 of the European Convention on Human Rights, are not violated if a local planning authority refuse to accord them a public inquiry, or, indeed, any form of oral hearing. This is because a combination of the authority's

initial decision-making process and judicial review by the High Court is sufficient to ensure compliance with Article 6. A

[47] Accordingly, issue (ii) is moot. Nevertheless, I wish to add a few words of my own on this issue, since it raises an important question as to the nature and scope of the power accorded to the Secretary of State by section 77 of the Act. Simon Brown LJ has set out the relevant provisions at [1]. I draw attention to two points. First, the section does not specify any B criteria for the exercise of the power. The Secretary of State is given an apparently unfettered discretion (subsection (1)) to give directions requiring applications for planning permission to be referred to him, instead of being dealt with by local planning authorities, and (subsection (2)) to give a direction either to a particular local planning authority, or to local planning authorities generally, that may relate to a particular C application or to applications of a specified class. Second, the Secretary of State can exercise the power at any time before the planning authority have "dealt with" the application.

[48] Mr McCracken submits that the Secretary of State is under a duty to exercise his section 77 power so as to prevent violations of Article 6. In my judgment, this submission is misconceived. He relies D strongly upon a single sentence in para 159 of the speech of Lord Clyde in *R (on the application of Alconbury Developments Ltd)* v *Secretary of State for the Environment, Transport and the Regions* [2001] UKHL 23; [2001] 2 WLR 1389, which has been quoted by Simon Brown LJ at [34]. But, properly understood, that sentence provides no support for Mr McCracken's argument. It is necessary to cite a little more from the E passage in question:

> As I indicated at the outset, Parliament, democratically elected, has entrusted the making of planning decisions to local authorities and to the Secretary of State with a general power of supervision and control in the latter. Thereby it is intended that some overall coherence and uniformity in national F planning can be achieved in the public interest and that major decisions can be taken by a minister answerable to Parliament. Planning matters are essentially matters of policy and expediency, not of law. They are primarily matters for the executive and not for the courts to determine

Lord Clyde was referring back, in particular, to paras 140 and 141 of his speech. At para 140, he had said: G

> Planning and the development of land are matters which concern the community as a whole, not only the locality where the particular case arises. They involve wider social and economic interests, considerations which are properly to be subject to a central supervision. By means of a central authority some degree of coherence and consistency in the development of land can be secured. H

It can be seen, therefore, that, when he spoke of "supervision and control" of local planning authorities by the Secretary of State, Lord Clyde was not referring to a function analogous to the supervisory jurisdiction exercised by the court when determining, for example, an application to quash the grant of a planning permission. He was referring to something very different, namely a planning function exercisable by the Secretary of State on planning merits, having regard to the public interest and the policy considerations identified by Lord Clyde.

[49] Accordingly, the passage in Lord Clyde's speech relied upon by Mr McCracken affords no support for his argument. Quite the contrary. I would add that the fact that the exercise of the section 77 power is an alternative to an application for planning permission being "dealt with" by the local planning authority shows that it was not the intention of parliament that the function of the Secretary of State should be to make good any shortcomings in the process undertaken by the planning authority. Parliament intended that applications for planning permission would usually be dealt with at local level by local planning authorities; but that, exceptionally, they could be dealt with by the Secretary of State if he decided to call them in. The two procedures are plainly alternatives. The purpose of the power conferred by section 77 is as described by Lord Clyde. It is to give the Secretary of State the power to call in planning applications where he considers that this is necessary or desirable in the national interest. It is not to exercise some supervisory control over the process by which local planning authorities perform their functions in individual cases. That is the function of the courts. I would endorse the passages in the first instance judgments to which Simon Brown LJ has referred on this point.

[50] Mr McCracken counters the argument, that it is for the courts to supervise the local planning authorities, by submitting that this argument leaves out of account the well-established principle that, save in exceptional cases, parties should not have recourse to the High Court where they have an adequate alternative remedy. The principle is indeed well established. But Mr McCracken's riposte begs the question of whether the section 77 power does provide an alternative remedy to judicial review. For the reasons that I have summarised, as well as those stated by Simon Brown LJ, the section 77 power does not provide an alternative remedy to judicial review. It is not the function of the Secretary of State to exercise his power under section 77 in order to make good any procedural shortcomings in the process adopted, or to be adopted, by the local planning authority.

[51] I too would dismiss this appeal.

Appeal dismissed.

Rugby Football Union *v* Secretary of State for Transport, Local Government and the Regions and another

[2002] EWCA Civ 1169

COURT OF APPEAL
Schiemann and Longmore LJJ and Ferris J
17 July 2002

Certificate of lawfulness of proposed use – Whether use of rugby A
ground for concerts within Class D2(b) and/or Class D2(e) of Town and Country Planning (Use Classes) Order 1987

The appellant applied to the local planning authority for a certificate of lawfulness of proposed user to use a rugby ground as a concert venue. The Secretary of State dismissed the appellant's appeal against B
non-determination of that application within the required eight-week period. The court below dismissed an appeal against that decision. The appellant appealed, contending that the Secretary of State had erred in concluding that: (i) the proposed use as a concert venue fell outside the scope of "use as a concert hall" within Class D2(b) of the Town and Country Planning (Use Classes) Order 1987 (UCO); alternatively (ii) the C
proposed use fell outside the scope of "an area for other outdoor sports or recreations" within Class D2(e).

Held The appeal was dismissed.

The function of the UCO is to enable issues that are unlikely to raise substantial planning issues to slip through without control. The correct way of approaching the UCO is to establish the use class within which the D
proposed activity is to be found – here Class D2 – and then to determine whether that activity can be found in the same use class. The correct question to ask in the instant case was whether the appeal site was being used as a concert hall. The answer to that question was "no": see [4]. Concerts do not fall within Class D2(e) because a concert performance is neither a sport nor a recreation, nor can it be classified as watching a E
recreation. Where there is doubt as to whether the UCO should be construed widely or restrictively, the approach should be governed by section 55(1) of the Town and Country Planning Act 1990, which defines "development". The 1990 Act, expressed as it is, effectively provides an exception to the general principle that any material change of use of buildings or land constitutes development. It follows ordinary principles of statutory F
interpretation, so that if a landowner wishes to rely upon the UCO, the burden of proof must be upon him to show that he comes within the order: see [16].

Cases referred to in the judgments

Forkhurst Ltd v *Secretary of State for the Environment* (1983) 46 P&CR 89; [1982] JPL 448, QB

A *Millington* v *Secretary of State for the Environment, Transport and the Regions* [1999] 3 PLR 118; [2000] JPL 297, CA

Nimmo v *Alexander Cowan & Sons* [1968] AC 107; [1967] 3 WLR 1169; [1967] 3 All ER 187, HL

Tessier v *Secretary of State for the Environment* [1976] 74 LGR 279; (1975) 31 P&CR 161

B **Appeal against a decision made under section 288 of the Town and Country Planning Act 1990**

This was an appeal by the appellant, Rugby Football Union, against a decision of Ouseley J, dismissing its appeal against a decision by the respondent, the Secretary of State for Transport, Local Government and the Regions, who dismissed the appellant's appeal against a
C non-determination, under section 195 of the Town and Country Planning Act 1990, by Richmond upon Thames London Borough Council.

Christopher Lockhart-Mummery QC and Thomas Hill (instructed by Denton Wilde Sapte) appeared for the appellant, Rugby Football Union.

Timothy Mould (instructed by the Treasury Solicitor) represented the first
D *respondent, the Secretary of State for Transport, Local Government and the Regions.*

The second respondents, Richmond upon Thames London Borough Council, did not appear and were not represented.

The following judgments were delivered.

E **SCHIEMANN LJ:**

[1] Before the court is an appeal by Rugby Football Union from a judgment of Ouseley J in a planning matter. The judge had upheld the finding of an inspector appointed by the Secretary of State. The background is usefully set out in paras 2 to 9 of the judgment, which reads as follows:

F

> 2. The certificate was sought in respect of an unlimited number of concerts. The Council failed to determine the application within the normal eight week period. On 11th July 2000 an appeal was lodged against that non-determination pursuant to section 195 of the Act. The appeal was dismissed by the Secretary of State's appointed person in a decision letter dated 19th January 2001. The RFU under section 288 of the 1990 Act now
> G challenges his decision.
>
> *THE LAW*
>
> 3. Section 192 of the Town and Country Planning Act 1990 provides:
>
> "(1) If any person wishes to ascertain whether –
>
> (a) any proposed use of buildings or other land; or
>
> ...
>
> H would be lawful, he may make an application for the purpose to the local planning authority specifying the land and describing the use or operations in question.
>
> (2) If, on an application under this section, the local planning authority are provided with information satisfying them that the use or operations described in the application would be lawful if instituted or begun at the

time of the application, they shall issue a certificate to that effect; and in any other case they shall refuse the application." A

4. A proposed use will be lawful if, among other things, it does not amount to "development" within the meaning of section 55 of the 1990 Act, since in those circumstances permission will not be required for it by virtue of section 57 of the Act.

5. Section 55(2) of the Act provides:

"The following operations or uses of land shall not be taken for the purposes of this Act to involve development of the land – B

...

(f) in the case of buildings or other land which are used for a purpose of any class specified in an order made by the Secretary of State under this section, the use of the buildings or other land or, subject to the provisions of the order, of any part of the buildings or the other land, for any other purpose of the same class;" C

6. Article 3(1) of the Use Classes Order provides that:

"Subject to the provisions of this Order, where a building or other land is used for a purpose of any class specified in the Schedule, the use of that building or that other land for any other purpose of the same class shall not be taken to involve development on the land."

7. Class D2 of the Schedule to the Use Classes Order includes the following uses: D

"**Assembly and leisure**

Class D2

Use as –

(a) a cinema,

(b) a concert hall, E

(c) a bingo hall or casino,

(d) a dance hall,

(e) a swimming bath, skating rink, gymnasium or area for other indoor or outdoor sports or recreations, not involving motorised vehicles or firearms."

THE FACTS

8. It was agreed before the appointed person by the London Borough of Richmond upon Thames and before me by the Secretary of State that the Twickenham stadium was within Class D2(e) of the Schedule to the Use Classes Order, although as the argument developed before me, it became clear that there was some disagreement as to the basis upon which it was there. It was however at least within Class D2(e) as "an area for other outdoor sports". F

9. The RUF submitted that the Secretary of State had erred in law in concluding that the proposed use as a concert venue fell outside the scope of "use as a concert hall" within Class D2(b) of the Use Classes Order and, alternatively, erred in law in concluding that the proposed use fell outside the scope of "an area for other outdoor sports or recreations". G

[2] Two issues arise on this appeal. The first was termed the "concert hall issue" by the judge. He said, at para 13: H

On the first issue, which was whether the proposed use would fall within Class D2(b) of the Schedule to the Use Classes Order as a concert hall, the appointed person said this:

"Taking first the appellants' argument that the proposal would amount to use as a concert hall, in Class D2(b), I accept that the UCO as a whole can

> A apply both to buildings and to other land; and I am also aware that the categories in Class D2(a) to (d) do not refer primarily to types of buildings but to types of use. Nevertheless, the use in Class D2(b) is defined specifically as "use as a concert hall", not more generally as use for holding concerts or as a concert venue, and it seems to me that, having regard to the ordinary meaning of the words, a use as a concert hall can only take place in a location having characteristics which enable it to function in the nature of a hall. The
> B dictionary definition of "hall" is a large room or chamber. I therefore tend to agree with the Council's view that use as a hall of any sort must necessarily be restricted to a building or part of a building that is enclosed both vertically and horizontally. For example, a building designed for use as a theatre, cinema or assembly hall could, in my view, function as a concert hall with little or no adaptation; similarly, any large enclosed building, such as a
> C warehouse or aircraft hangar, could have characteristics enabling it to be used as a concert hall, within the scope of Class D2(b).
>
> 6. By contrast, a mainly open stadium like Twickenham does not, in my view, possess the necessary degree of enclosure to function as a concert hall, and I do not consider that the inclusion of both indoor and outdoor activities in Class D2 as a whole is sufficient to extend the use defined very precisely in Class D2(b) to that extent."

D

In para 20, the judge goes on:

> the question is: for the purposes of the Use Classes Order, is it sufficient to focus on the purpose or use to which concert halls are put, ie holding concerts? Or is it necessary to focus on the question of whether Twickenham has the physical characteristics of a concert hall, when concerts take place
> E there?
>
> 21. ...The natural structure of the Use Classes Order means that in order to qualify for the change from a use within D2(e) to the use within D2(b), the change from "use as an area for other outdoor sports" to "use as a concert hall", the reference to "use for a purpose of the same class" within section 55 and article 3 of the Use Classes Order simply requires that the language of the two applicable paragraphs be satisfied. Those are the purposes in this
> F context. The appointed person, therefore, correctly asked the question: when concerts take place there, is Twickenham used as a concert hall? To which he correctly answered that it was not, because it lacked an essential physical characteristic – it had no roof. It is insufficient for it simply to be a concert venue. The building is not and never could be used as a concert hall because it lacks the essential characteristic for a concert hall, namely a roof.
>
> G 22. I consider that in order to be able to take advantage of the changes within the Use Classes Order it is necessary for the essential physical characteristics of (in this case) a roof to be present in order for the change of use to be exploited...
>
> 23. I accept that some classes or purposes within the Use Classes Order use the different phraseology of, for example, in Class D1 "use for the provision...", or within Class A1 under the rubric of "shop":
>
> H "Use for all or any of the following purposes –
>
> (a) for the retail of goods...
>
> (b) as a post office."
>
> 24. That distinction, however, reinforces my view that one looks simply to see if the requirement of the particular paragraph is satisfied: is this stadium, therefore, used as a concert hall? The question is not: is this a building which

> although not a concert hall is nonetheless being used for the purposes for which a concert hall is used? The Use Classes Order would simply have referred to use "for concerts" or "as a concert venue".
>
> 26. I do not consider that this approach deprives the Use Classes Order of effect in relation to changes from outdoor sports. The Use Classes Order permits the change of sport where it might be thought that the change of sport involved a different use because there could be different impacts from the relative popularity of the sports involved, or from degrees of professionalism or, for example, from the use of an area for training as opposed to playing matches.

[3] Mr Christopher Lockhart-Mummery QC attacked this judgment as follows. He took us through the history of the Use Classes Order (UCO) since 1948, and stressed that the wording reflected contemporary perceptions as to what could be allowed to escape the planning approval process that otherwise might have required such approval. He emphasised that planning is concerned with land use. He submitted that, in planning law, it is always necessary to identify the purpose of a land use in order to identify its status. He submitted that, therefore, both in section 54(2)(f) and in article 3 of the UCO, one must answer the question "for what purpose will the appeal site be used after the proposed change of use?". He submitted that when one asks that question in the context of the present case, the rugby football ground will be used for the purpose of holding concerts there.

[4] It is a simple and, as the judge recognised in giving leave, an arguable point. Like the judge, I think it wrong, essentially for the reasons that he gave. The function of the UCO is to enable things unlikely to raise substantial planning issues to slip through without control. The simplest and, in my judgment, the correct way of approaching the UCO, in a case like the present, is to establish the use class within which the current activity is to be found – in the present case, Class D2 – and then to see whether the proposed activity can be found in the same use class. One should assume that the proposed activities are taking place and ask oneself: "Is the appeal site being used as a concert hall?" That is the proper question to ask. In the present case, the natural answer to that question is "no".

[5] Like the judge, I fully accept that some other classes of the UCO pose the question in a different form, and it will therefore become necessary to ask the question in the form set out in the relevant use class. The questions are different in form as well as in content. There are two main types of formulation of the question to be found in the order. I give three examples of each. Thus, Class A1 poses the question: "Is the land to be used for the retail sale of goods?" Class B1(c) poses the question: "Is the land to be used for an industrial process?" Class D1(c): "is the land to be used for the provision of education?" On the other hand, Class C1 poses the question: "Is the land used as an hotel?" Class D1(f): "Is the land used as a public library or public reading room?" And Class D2(b) poses the question: "Is the land used as a concert hall?" So far as that latter category of three questions is concerned, the fact that someone makes his field

A available for overnight tenting does not cause the land to be used as an hotel; the fact that someone is good enough to display his books in a field does not mean that the field is used as a public reading room; and the fact that someone is good enough to let his field be used for a band giving military concerts does not cause that land to be used as a concert hall.

[6] Like the judge, I consider that the fact that the UCO is not consistent B in the way in which it defines use classes is a point against the appellant.

[7] I turn now to the second issue, which the judge headed "Outdoor Sport and Recreations". He said, in [31]:

> The second issue was whether Twickenham stadium as at least an area used "for other outdoor sports" (ie rugby) would be used as a venue "for other outdoor sports or recreation" when used for concerts.
>
> C 32. The appointed person said:
>
> "The appellants submitted that the public enjoyment of concerts constituted a recreation. I agree that it is certainly a recreation, but the issue here, as they recognise, is whether it is a recreation within the context of Class D2(e). The use of the term 'other… recreations' (my emphasis) [not the appointed person's emphasis] indicates to me that the scope of Class D2(e) only extends D to recreations which are of a similar nature to those which might take place in a swimming bath, skating rink or gymnasium. Otherwise, use as a swimming bath, skating rink or gymnasium might be expected to have been included in separate sub-paragraphs of the Class, like each of the uses in paragraphs (a) to (d), and use as an area for other indoor or outdoor sports or recreations would not have been included with these specific activities."

E [8] There are then further quotations from the inspector.

[9] The judge says, in [42]:

> I did not find this issue straightforward. My mind changed a number of times. However, I have come to the firm conclusion that… The focus of Class D2(e) is on sport or physical recreation. The words "or other outdoor recreation" need to be read in the context of Class D(2)(e) which is clearly F dealing with physical activities rather than hobbies, interests, or recreational activities of an artistic or creative nature. It cannot cover all those ways in which a person can enjoy recreation in a broad sense without becoming so broad as to render the rest of Class D2 otiose, or creating quite remarkable possibilities for uncontrolled changes of use. It would be inappropriate so to interpret the extension of the Use Classes Order in 1987.
>
> 43. I also conclude that the words in question should be seen as relating to G that group of activities which it extends, which are those in Class D2(e), rather than to all those in Class D2 as a whole. The connotation and context of D2(e) is sport and physical recreation, the role of physical recreation being to cover those many situations where the physical activity would not be described as a sport because, for example, it might be done alone for private pleasure, I instance for these purposes a jogging track. This approach fits with the approach of the Court of Appeal in *Millington* v *Secretary of State for* H *Environment, Transport and the Regions* and *Shrewsbury* v *Atcham Borough Council* [2000] JPL 297.
>
> 44. I do not consider therefore that the presence of spectators is material to Class D2(e). Twickenham is within Class D2(e) because it is an area used for outdoor sport, in this instance rugby union. It is not the presence of more or less passive spectators which brings it within the purpose of Class D2(e).

> They may enjoy watching rugby. For them, in common sense language, watching rugby may be one of their recreations but they are not themselves engaged in physical recreation. Twickenham is not, for Class D2(e) purposes, an area used for outdoor recreation by spectators.
>
> 45. So far as concert performers are concerned it is only reasonable to assume that those contemplated here are professional musicians. No doubt they derive much pleasure from their own music-making but they are not for that reason alone engaged in recreation. Nor are they engaged in physical recreation however much the performance may make demands on their stamina. Their activities do not bring Twickenham within the scope of other outdoor recreation.
>
> 46. The concert goers are the equivalent to the rugby spectators for these purposes, but their enjoyment of the music or spectacle does not constitute "other recreation" for the purposes of Class D2(e) any more than does the rugby spectators' enjoyment of rugby. The reason why Twickenham stadium is within Class D2(e) is because sport is *played* there [my emphasis] and not because sport is *watched* [my emphasis] there. The reason why the concerts are not within Class D2(e) is that the concert performance is neither sport nor recreation, nor is watching the recreation.
>
> 47. The distinction between watching and playing sport means of course that an important aspect of Twickenham stadium is disregarded. However, the distinction is a perfectly rational one in the context of the Use Classes Order. It is perfectly rational to focus on physical activity rather than on watching physical activity. If Twickenham had to be approached as an area used for watching professional sport it would not be within Class D2(e) at all.

[10] I ought to add something in relation to *Millington* v *Secretary of State for the Environment, Transport and the Regions* [2000] JPL 297[1] where I delivered the only substantive judgment. One of the issues in that case was whether the landowner could rely upon the UCO in defence of an enforcement notice that charged him with using his land for the purposes of selling wine and light refreshments, and for visits of fee-paying members of the public. In order to advance this argument, he needed to establish that his previous use fell within Class D2 and was the use of an area for recreations. The previous use consisted of visits by members of the public to look at the remains of the old Roman town of Viriconium in Shropshire. The inspector held that such visits may have had a recreational component, but, having heard that a good number of the visitors were archeologists and students, the main purpose of the visit was probably out of academic interest. The Secretary of State agreed with the inspector that he did not regard the presence of an ancient monument on the land, and the particular interest shown therein by visitors, as a recreational use. This court held that the Secretary of State was entitled to take that approach. I said, at p305, that the Secretary of State was perfectly justified in finding that:

[1] [1999] 3 PLR 118

A in the context of the use of the words in the Use Classes Order, what had gone on in the days before the vineyard was established did not amount to the use of the land for the purposes of recreation.

[11] Mr Lockhart-Mummery criticised the approach of the judge on two main grounds: (1) an *eiusdem generis* approach was not an appropriate approach to construing the UCO that had grown up over the years. I have B some sympathy with the argument, but would reject it. The *eiusdem generis* approach certainly appealed to the Court of Appeal in *Millington*; (2) the judge erred in incorporating some physical requirement into the concept of recreation in Class D2. Mr Lockhart-Mummery submitted that it would have been easy for the legislator to have inserted such an adjective, and yet he did not do so. He submitted that a much wider meaning could be given C to "recreation", so as to embrace bridge tournaments, stamp collecting, chess competitions and the like. I accept that the word is capable of having wider meanings. So did the judge. I accept that the general heading of Class D is "assembly and leisure", which undoubtedly covers some of the D2 activities, such as use as a bingo hall. But I agree with the judge that, in the present context, which is D2(e), it is right to incorporate some physical D effort to the phrase "recreation", otherwise it is so wide as to be an improbable planning tool. Mr Lockhart-Mummery, in his very helpful skeleton argument, referred the court to the definition of "recreation" in the *Oxford English Dictionary* under heading 3a, which is as follows:

> The action of recreating (oneself or another), or fact of being recreated, by E some pleasant occupation, pastime or amusement.

[12] I noticed, on looking at the source, that among the examples that the learned editor cites in support of that definition is the following from Hook "Lives AGPS" in the last century: "He sought his recreation in the study of Ecclesiastical history." All we can get from the *Oxford English Dictionary* is that the word is capable of having a very wide meaning. For F my part, I consider, and I think probably Mr Lockhart-Mummery was constrained to agree, that the drafter of the UCO can hardly have had that type of recreation in mind.

[13] If one gives recreation as wide a meaning as that for which Mr Lockhart-Mummery contends, then the whole of the rest of Class D2 is otiose. I see some force in Mr Lockhart-Mummery's point that if one G inserts "physical", as the judge did, then the rest of Class D2(e) is surplus to requirement. However, that seems to me to be a smaller point, and arguably explicable on the basis of the legislative history and a desire not to wish to alter things. But perhaps the points do balance one another.

[14] Essentially, for the reasons given by the judge, I would reject Mr Lockhart-Mummury's arguments and dismiss this appeal.

H

LONGMORE LJ:

[15] I agree with my lord, for the reasons he gives, that this appeal must be dismissed. There was some discussion in argument on the question of whether the classes of use, as set out in the Town and Country Planning (Use Classes) Order 1987 should be construed widely or restrictively. We

were referred to an apparent difference of opinion on that matter between Lord Widgery CJ, in *Tessier* v *Secretary of State for the Environment* (1976) 31 P&CR 161, favouring a restrictive view, and Hodgson J in *Forkhurst Ltd* v *Secretary of State for the Environment* [1982] JPL 448, favouring a neutral view. That difference of view is not important in this case, since it is clear, by ordinary processes of construction, that the appeal has to fail. In a case of genuine doubt, however, my own approach would be governed by the words of section 55(1) of the Town and Country Planning Act 1990, which defines "development" as, *inter alia*, "the making of any material change in the use of any buildings or other land". Section 55(2) then instances various operations or uses of land that are not to be taken to involve development of the land. One of those instances is contained in section 55(2)(f): A B C

> in the case of buildings or other land which are used for a purpose of any class specified in an order made by the Secretary of State under this section, the use of the buildings or other land or, subject to the provisions of the order, of any part of the buildings or the other land, for any other purpose of the same class;

D

[16] It seems to me that this Act of parliament, expressed in the way that it is, is effectively providing an exception to the general principle that any material change in the use of buildings or land is to constitute development. It follows, upon ordinary principles of statutory interpretation, that if a landowner wishes to rely upon the UCO, the burden of proof must be upon him to show that he comes within the order: see, in general, *Nimmo* v *Alexander Cowan & Sons Ltd* [1968] AC 107, at pp129-130 *per* Lord Wilberforce, in a passage from his dissenting speech that was expressly approved by the majority of the House of Lords in *R* v *Hunt* [1987] AC 352. E

[17] On any view, however, the claimant, as here, fails to show that he comes within the order, and, in my judgment, the appeal must fail. F

FERRIS J:

[18] I agree with both judgments and do not wish to add anything.

Appeal dismissed.

JJ Gallagher Ltd *v* Secretary of State for Transport, Local Government and the Regions and another

[2002] EWHC 1812 (Admin)

QUEEN'S BENCH DIVISION: ADMINISTRATIVE COURT
Mr George Bartlett QC (sitting as deputy judge of the Queen's Bench Division)
23 August 2002

PPG 6 "Town Centres and Retail Developments" – Sequential approach – Secretary of State disagreeing with inspector's recommendation – Whether Secretary of State correctly applied sequential approach – Whether Secretary of State gave adequate reasons for rejecting inspector's conclusions

The claimant had applied for planning permission to develop a non-food retail park (Class A1 use) in respect of a 10ha site. In February 2000, planning permission had been granted, on appeal, for Class A1 retail development of part of the site (the Church Commissioners site). In February 2001, an inspector, who had held an inquiry into the appeal in respect of the larger site, recommended that planning permission be granted. The Secretary of State requested that the inspector should provide two addenda to her report. In November 2001, and against the recommendations of his inspector, the Secretary of State dismissed the claimant's appeal, concluding that the inspector had misinterpreted part of the relevant statutory planning policy and had wrongly applied the sequential approach to site selection when there were alternative suitable sites available. No reference was made to his earlier decision on the Church Commissioners site. The claimant challenged that decision on the grounds that the Secretary for State had: (i) adopted an unfair process in reaching his decision, and had failed to comply with the requirements of r 17(5) of the Town and Country Planning (Inquiries Procedure) Rules 2000; (ii) misinterpreted his own policy on retail development; and (iii) applied the sequential approach incorrectly.

Held The application was granted.

The first ground of challenge failed. In no respect were the claimant's interests adversely affected by the procedure adopted by the Secretary of State: see [31]. The Secretary of State did not misinterpret his own policy on retail development. The passage in the government's response that contrasts the "format-driven" and the "class-of-goods" approaches, and provides that it is the latter that accords with PPG 6, is not to be read as confined to food retail policy. The "class-of-goods" approach focuses on the range of goods proposed to be sold, and considers whether such goods could not be sold in a town-centre development, if necessary of a format, design and scale different from the development proposed. There was no inconsistency between this approach, which the Secretary of State evidently treated as being implicit in PPG 6, and the express words that were used: see [34]. The Secretary of State did apply the sequential test incorrectly. He did not say what weight, if any, he attached to the demand for large retail warehouses on the part of a sector of the retail market: see [48]. He failed to explain his views on the suggested unattractiveness of alternative sites to retail investors: see [49]. He failed to say why he rejected the inspector's conclusion that certain of the alternatives ceased to be contenders on account of their location. There was a need for the Secretary of State to explain why he based his decision upon the council's view of the date upon which capacity would arise, rather than upon the claimant's view, derived from a specific study: see [54]. The Secretary of State should have explained why he was distinguishing the present appeal from the permission given on appeal in respect of the Church Commissioners site: see [56]. The Secretary of State did not take account of the claimant's supplementary evidence on the sequential test: see [59].

Cases referred to in the judgment

Luke (Lord) of Pavenham v *Minister of Housing and Local Government* [1968] 1 QB 172; [1967] 3 WLR 801; [1967] 2 All ER 1066; (1967) 65 LGR 393; 18 P&CR 333; 202 EG 1121, CA

North Wiltshire District Council v *Secretary of State for the Environment* (1992) 65 P&CR 137; [1992] 3 PLR 113; [1992] JPL 955, CA

R v *Chief Constable of Thames Valley, ex parte Cotton* [1990] IRLR 344, CA

Application under section 288 of the Town and Country Planning Act 1990

This was an application, under section 288 of the Town and Country Planning Act 1990, by the claimant, JJ Gallagher Ltd, seeking to quash a decision of the first defendant, the Secretary of State for Transport, Local Government, and the Regions, dismissing an appeal against a refusal of planning permission by the second defendants, Gateshead Metropolitan Borough Council.

Ian Dove and Christopher Young (instructed by The Wood Glaister Partnership, of Solihull) appeared for the claimant, JJ Gallagher Ltd.

Timothy Mould (instructed by the Treasury Solicitor) represented the first defendant, the Secretary of State for Transport, Local Government, and the Regions.

The second defendants, Gateshead Metropolitan Borough Council, did not appear and were not represented.

The following judgment was delivered.

MR GEORGE BARTLETT QC:

[1] In this application, under section 288 of the Town and Country Planning Act 1990, the claimant seeks to have quashed a decision of the first defendant, dismissing an appeal against a refusal of planning permission by the second defendant. The planning application sought permission for development consisting of a non-food retail park

A (Use Class A1), including access, parking, landscaping and associated highway works on a 10ha site (the appeal site) at Harrington works, at the western end of the MetroCentre shopping complex in Gateshead. On 23 February 2000, planning permission had been granted on appeal for Class A1 retail development on 3.94ha of the appeal site (the Church Commissioners site). An inspector held an inquiry into the B appeal on the appeal site over four days in January 2001, and, in her report of 16 February 2001, recommended that planning permission should be granted. The Secretary of State requested from the inspector two addenda to her report, and the inspector submitted these on 19 March 2001 and 17 July 2001. In his decision of 22 November 2001, the Secretary of State rejected the inspector's recommendation that C planning permission should be granted. He did so because he considered that the inspector had misinterpreted part of the relevant statutory planning policy, and because he concluded that the sequential approach to site selection required by central government policy had not been satisfactorily carried out and that there were alternative suitable sequentially preferable sites available. He did not D refer in his decision letter to his earlier decision on the Church Commissioners site.

[2] The challenge to the Secretary of State's decision is made on three principal grounds. The first relates to the post-inquiry procedures, in the course of which the Secretary of State received the two addenda to the inspector's report. It is said that these procedures were unfair. Second, it is E said that he misinterpreted his own policy in relation to the sequential approach. The third principal ground relates to the Secretary of State's conclusion that the appeal scheme failed to accord with the sequential approach. Although originally couched in terms of a perversity contention, this third ground resolved itself principally into a reasons challenge, with an additional argument that certain material F considerations (one of which was the recent decision in the Church Commissioners appeal) had been left out of account.

Planning policy

[3] The statutory development plan that the inspector and the Secretary of State were required to take into account was the Gateshead unitary G development plan (UDP), which had been adopted in November 1998. Policy S2 of the UDP identifies the MetroCentre as an established regional shopping centre, and says that support will be given to the continuation of this role where it does not undermine the vitality and viability of existing centres, and where it maximises the opportunity to use means of transport other than the car. Policy S3 provides that, within the MetroCentre H boundary, the development of retailing and appropriate employment uses will be permitted where it can be demonstrated that: they accord with the development plan strategy; do not undermine the vitality and viability of existing centres; contribute towards achieving improved accessibility by public transport, by cycling and by walking; and further a reduction in dependence upon the private car.

[4] Paragraph 7.13 of the UDP is the provision that gave rise to the disagreement between the inspector and the Secretary of State. It provides:

> All proposals for retail development over 2,500 square metres gross floorspace will need to be supported by evidence on these matters in the form of impact assessments, including consideration of the cumulative effects of recently completed and proposed developments. It should also be demonstrated that a sequential approach has been adopted to site selection, whereby priority has been given to locations within or on the edge of town centres. If no suitable sites are available in such locations, development within the MetroCentre boundary is the preferred location for new retail development within the Borough.

[5] In para 71 of her report, the inspector, having referred to this provision of the UDP, said:

> In my view this supporting text relating to the sequential test is secondary to policy as the Plan identifies the MetroCentre as the preferred retail location within the Borough.

[6] The inspector then concluded:

> 74. In policy terms, therefore, retail development is appropriate on this site and the appeal site is a preferred location for retail warehouse development. The proposals are therefore in accordance with UDP policy.

[7] The Secretary of State disagreed with this. He said:

> 14. The Secretary of State does not agree with the Inspector's assertion (IR71) that paragraph 7.13 relating to the sequential test is secondary to policy as the Plan identifies the MetroCentre as the preferred retail location within the Borough. Policy S2 of the UDP describes the MetroCentre as "an established regional shopping centre", but does not refer to it as the preferred retail location. Policy S3 sets out a series of tests which must be fulfilled before certain types of planning application at the MetroCentre will be permitted. There is nothing in the policies themselves, nor in the reasoned justification, to indicate that the MetroCentre is the preferred retail location. Furthermore, the UDP does not define the MetroCentre as a city, town, district or local centre. The Secretary of State therefore considers that in terms of the sequence of preferences set out in paragraph 1.11 of PPG 6, it is the least preferable location for new retail development.

[8] The claimant does not seek to argue that the Secretary of State erred in disagreeing with the inspector on her conclusion that para 7.13 was secondary to policy, and that UDP policy identified the MetroCentre as the preferred retail location for the borough.

[9] PPG 6 (Planning Policy Guidance Note 6, Town Centres and Retail Developments, issued in June 1996) sets out the Secretary of State's policy on the sequential approach. It says:

> 1.11 Adopting a sequential approach means that first preference should be for town centre sites, where suitable sites or buildings suitable for conversion are available, followed by edge-of-centre sites, district and

> local centres and only then out-of-centre sites in locations that are accessible by a choice of means of transport.
>
> 1.12 The Government recognises that the approach requires flexibility and realism from local planning authorities, developers and retailers. Developers and retailers will need to be more flexible about the format, design and scale of the development, and the amount of car parking, tailing these to fit the local circumstances. Local planning authorities should be sensitive to the needs of retailers and other town centre businesses and identify, in consultation with the private sector, sites that are suitable, viable for the proposed use and likely to become available within a reasonable period of time.

[10] Paragraph 3.2 of PPG 6 states that the sequential approach applies both to comparison shopping and to convenience shopping. Paragraph 3.3 recognises that some types of retailing, such as large stores selling bulky goods, may not be able to find suitable sites either in or on the edge of town centres. In such cases, it says, the local planning authority should seek to ensure that such developments are located where they will be easily accessible by a choice of means of transport.

[11] In 2000, the House of Commons Select Committee on the Environment, Transport and Regional Affairs produced its second report – *Environmental Impact of Supermarket Competition*. In May 2000, the Department of the Environment, Transport and the Regions published the government's response to this report. Certain paragraphs in the response, which are relevant to the claimant's second ground of challenge, need to be noted:

> 5. The Government agrees that there should be a consistent interpretation of the sequential test. Para 1.12 of PPG 6 states clearly that the Government expects "developers and retailers will need to be more flexible about the format, design and scale of the development, and the amount of car parking, tailoring these to fit the local circumstances". The Government has a clear policy that new retail investment should be used to strengthen existing centres and expects this paragraph of PPG 6 to be read in that light.
>
> 6. Research recently published by the National Retail Planning Forum, and part funded by the Department of the Environment, Transport and the Regions, considered how the sequential approach is being applied in practice and its effects on retail development. This showed an inconsistency in the application of the sequential approach and contrasts two approaches taken:
>
> (i) the format-driven approach, where the developer has a preferred format, which, in their view, does not fit into town centres and should be exempted from the policy; or
>
> (ii) the "class of goods" approach, where the key question is whether there is any reason why such goods cannot be sold from town centres.
>
> 7. The Government's policy – as stated in PPG 6 – is to take the latter approach, as most goods can be sold from town centres. PPG 6 does not propose exceptions to the application of the sequential approach.
>
> 8. The Select Committee has suggested that there is a need to clarify how the policy should apply to large-format developments. The Government considers that PPG 6 is clear – it expects developers to demonstrate flexibility. It recognises that the need to demonstrate flexibility will have consequences for retailers who insist on the need for very large formats.

[12] The Secretary of State, in a passage of his decision letter that I will set out in due course, referred to PPG 6 and the second report as stating that it was government policy to favour the class-of-goods approach. The claimant's contention is that PPG 6 does not favour the class-of-goods approach, and that the second report applies only to food retailing. The argument that, accordingly, the Secretary of State misapplied his own policy forms the second ground of challenge. A B

Church Commissioners' appeal

[13] The decision of the Secretary of State, granting planning permission for non-food retail units on the Church Commissioners site, followed a public inquiry in November 1999 and a recommendation of the inspector, who held that permission should be granted. The inspector had concluded that the question of need had been determined by the UDP and did not have to be demonstrated. On the sequential test, he had had put before him a detailed sequential approach analysis. Referring to this, at para 45 of his report, he said: C

> Some 60 sites were investigated, and a further three once consultation responses on the planning application had been received. The Church Commissioners' conclusion that there are no suitable out-of-town or edge-of-town sites is accepted by GMBC. I have found no reason to take a different view. The UDP states that land within the MetroCentre boundary is the preferred location for retail warehouse development. D

[14] The conclusion that there were no sequentially preferable sites available was accepted by the Secretary of State. At para 2 of his decision letter he said that he agreed with all the inspector's conclusions, except the conclusion that need did not have to be addressed. On that point, he said: E

> The Secretary of State therefore considers that your client's proposals should be required to demonstrate need. He agrees with the Inspector (IR48) that, in this case, evidence was submitted to the inquiry by both your client and Gateshead MBC, which satisfactorily demonstrates that there is a need for the development in terms of retail capacity (in terms of the likely growth in retail expenditure); retailer demand; employment generation; development of a brownfield site; facilitating the development of adjoining land and improving the opportunities for access by means other than the private car. F

Post-inquiry procedures G

[15] In the present case, the claimant and the council were in agreement at the inquiry that need for the development had been demonstrated. The inspector recorded the claimant's case on this as follows:

> 36. With regard to the quantitative assessment, the Council's comments are based on the KS Revised Retail Impact Assessment of April 1999, rather than the GL Hearn study of November 2000. The Council therefore considers that there would be sufficient expenditure by 2006 to support all the 1999 commitments together with the appeal proposals. Whilst much of the KS methodology is accepted it significantly underestimated the capacity for additional comparison goods retailing because it did not take into account the capacity for additional floorspace which already existed in 1999 by its H

> A making allowance for an increase in the turnover of existing floorspace. On the GL Hearn analysis there would be sufficient capacity for the appeal scheme and other commitments by 2003 and there would still be a surplus of £123.10m to support additional development proposals or growth in the turnover of existing retailers. Indeed the level of growth is such that sufficient capacity to support the appeal scheme exists by the end of 2001. The need for the development proposed is therefore immediate. The Council B maintains that the need will not arise until 2006 (GMBC8) by which time all the current commitments, including the appeal proposals, could be comfortably supported. Indeed, the Council's witness accepts that by the time this appeal decision is issued the operational need will have arisen and the only dispute is where that need will be met.

C [16] In the light of this, under the heading "Need and impact", the inspector concluded as follows:

> 75. The need for the proposed development is not disputed. If those needs are met on the appeal site then the site specific, qualitative benefits relating to the redevelopment of a contaminated brownfield site and the improvement of public transport, footpath and cycle links to the D MetroCentre would be realised. Similarly, the Council does not argue matters of retail impact as it is agreed that the likely impact of the appeal proposals would not give rise to any unacceptably adverse effect on existing centres. (24, 25, 37, 33,38).

[17] The Secretary of State was not satisfied with the inspector's acceptance of the claimant's case on need. The decision letter said:

E

> 3. Having examined the Inspector's report, the Secretary of State was of the view that the Inspector had accepted without question (IR24 and 25) the agreement on the issues of need and impact of the development reached by your client and GMBC before the public inquiry began. There was no evidence that she had independently considered these matters, and her own conclusions on them are not recorded in the report (IR75).
>
> F 4. Accordingly, the Government Office (GO) requested an addendum to the Inspector's report from the Planning Inspectorate by telephone. The addendum was received in the Government Office on 19 March 2001. However, it was considered unsatisfactory, as it only contained a brief resumé of the events leading up to the agreement on need and impact, and the Inspector still did not offer a view on whether the agreement should be accepted.
>
> G 5. Following discussions with the Planning Inspectorate, and further detailed examination of the Inspector's report and the evidence before the inquiry, the Secretary of State requested a second addendum to the report by means of his letter of 21 June 2001. The letter set out in detail, in paragraphs 1(a) and 1(b), all the matters on which he wished to receive a view from the Inspector.

H [18] The inspector submitted her second addendum on 17 July 2001. She said:

> 6. Four retail impact assessments have been carried out by different consultants each taking a different approach to the assessment of need. The worst case postulated is that by 2006 expenditure growth will comfortably

support all the identified commitments. A less cautious calculation suggests that by 2003 there would be a residual of £36.7M even when confining the analysis to the 20 minute catchment area. This residual rises to £71.0M by 2006. Expenditure growth therefore suggests that there is capacity for the proposed development in terms of retail need…

8. In terms of qualitative need the appeal site shares the advantages identified by the Secretary of State in the smaller Gibside Way decision in relation to retailer demand, employment generation, development of a brownfield site, improving opportunities for access by means other than the private car. It also scores well in relation to access to low income areas and reducing the need to travel.

9. I conclude that the evidence before the inquiry satisfactorily demonstrates the existence of retail need.

[19] At para 17 of the decision letter, the Secretary of State said that, for the reasons given in these paragraphs of the inspector's second addendum, he agreed with the inspector's conclusions that the evidence before the inquiry satisfactorily demonstrated the existence of retail need.

[20] The letter of 21 June 2001, from Government Office North East (GONE) to the Planning Inspectorate, requesting the second addendum, was sent to the parties in letters dated 17 July 2001. It said that any representations that the party might wish to make in respect of the matters in the letter of 21 June should be sent to GONE within three weeks. The council replied on 3 August 2001. On 6 August 2001, the claimant's planning consultant, GL Hearn, replied, saying that it had found it impossible to make any meaningful comment without knowledge of the contents of the inspector's report. It requested a copy of the report. GONE refused to release the inspector's report, and, in a further letter of 16 August 2001, GL Hearn wrote in terms that effectively repeated its case on need at the inquiry.

First area of challenge: Post-inquiry procedure

[21] In its grounds of application, the claimant said that the process adopted by the Secretary of State in reaching his decision was unfair and in breach of Article 6 of the European Convention on Human Rights, in that he had failed to give the claimant a fair and public hearing, and, through refusing to disclose the inspector's report and its addenda, had prevented the claimant from making any sensible or informed comment. In addition, it was said that the Secretary of State had failed to comply with the requirements of r 17(5) of the Town and Country Planning (Inquiries Procedure) Rules 2000.

[22] The claimant sought disclosure of the correspondence, notes and documents that recorded, or referred to, telephone conversations between GONE and the Planning Inspectorate, and an order for disclosure was made. Particular reliance is placed by Mr Ian Dove upon correspondence and memoranda showing that, at one stage, GONE and the Planning Inspectorate had considered that a reopening of the inquiry would be necessary.

[23] Mr Dove's submission on the Town and Country Planning (Inquiries Procedure) Rules 2000 is that the Secretary of State had failed to

A comply with the requirements of r 17(5), in that he differed from the inspector on findings of fact, or, alternatively, that he had taken into account new evidence and matters of fact, and had failed to reopen the inquiry or to afford the claimant the opportunity to make representations to him. At paras 76 and 77 of her report, the inspector had said:

> B 76. The list of sites which the Council now considers to be sequentially preferable (GMBC13) was only finalised during the course of the inquiry and only those on "List 1" were seriously pursued. These could be considered to be sequentially preferable if they are able to perform the same or similar functions intended for the proposed development and within a reasonable timescale. A number of the sites suggested are not free from planning constraints and, unlike the appeal site, do not enjoy as favourable a UDP C allocation as the appeal site. Only 3 of the List 1 sites are suitable for retail warehouse development (40, 49).
>
> 77. Furthermore, the sequential test has to have regard to the catchment area which is intended to be served. If it is accepted that about 80% of the trade draw comes from within a 20 minute drive time then List 1 sites B2 (South Shields), B3 (Whitley Bay), B6 (Consett), C4 (Shields Road, Newcastle), and most of the sites on List 2 cease to be contenders. Also sites D which are in local or district centres, for example Consett B6, Killingworth B4, Cramlington C17 and Prudhoe C22, are not, in my view, capable of fulfilling the same function as the appeal proposals because, as paragraph 3.18 of PPG 6 states, retail development should be in scale with the centre in which it is to be located (48, 51).

[24] The Secretary of State, in a section of his decision letter that I will E refer to in more detail later, noted in para 27 that there were sites with planning permission for retail warehouses in Killingworth Town Centre (B4), Hermiston Retail Park (B6) and Shields Road District Centre (C4), and said that there was "no evidence that retail warehouse development of the scale already permitted on B4 and B6 would be out of scale with these centres". Later, in para 29, he said that sites B4, B6 and C4 were F suitable for retail warehousing use, and could be made available in the short-term. It is to these passages that, as I understand him, Mr Dove points as showing that the Secretary of State differed on findings of fact or, alternatively, took into account new factual material. As a consequence of these and other conclusions, the Secretary of State considered that there was no compelling need to provide the floorspace proposed.

G [25] Rule 17(5) applies where, after the close of the inquiry, the Secretary of State:

> (a) differs from the inspector on any matter of fact mentioned in, or appearing to him to be material to, a conclusion reached by the inspector; or
>
> (b) takes into consideration any new evidence or new matter of fact (not being a matter of government policy),
>
> H and is for that reason disposed to disagree with a recommendation made by the inspector…

[26] In my judgment, the matters upon which Mr Dove seeks to rely for his contention that there has been a failure to comply with reg 17(5) of the Inquiries Procedure Rules are properly to be regarded as matters of

planning judgment, rather than of fact. The need to draw such a distinction in applying the requirements of the Rules was made clear in relation to the predecessor provisions, in *Lord Luke of Pavenham* v *Minister of Housing and Local Government* [1968] 1 QB 172 (in particular at p191D-E, *per* Lord Denning MR). A

[27] Whether a site is suitable for retail warehousing development or whether a retail warehousing development would be out of scale with the centre within which it is proposed are matters of planning judgment. They are conclusions that are derived from factual matters – the physical nature of the site and its surroundings, the amount of floorspace in the centre, and so forth – but they are not themselves findings of fact. There is, moreover, nothing to suggest that, in reaching the conclusions that he did, the Secretary of State took into account any evidence or facts that were not before the inspector. As far as the inspector's supplementary reports are concerned, these did not refer to any new evidence or matter of fact, nor, in asking for them, was the Secretary of State doing any more than seeking conclusions from the inspector on factual matters that were already available to her, and had been available to the parties at the inquiry. In any event, it was not because of anything contained in the supplementary reports that the Secretary of State was disposed to disagree with the inspector's recommendation. The Secretary of State reached his conclusion, that permission should be refused in spite of, not because of, the inspector's conclusions, on retail need. Rule 17(5) was thus not engaged. B C D

[28] Apart from his specific reliance upon the requirement of r 17(5), Mr Dove puts his challenge to the Secretary of State's post-inquiry procedure on the more general grounds of fairness. He draws attention to para 3(ii) of the Ministerial Code, which was among the documents disclosed. This states: E

> In order to demonstrate even-handedness, and in the interests of natural justice, all evidence which is material to any decision which has been the subject of a planning inquiry, and which the decision taker ultimately takes into account, must be made available to all parties with an interest in the decision. F

[29] Mr Dove, as I have said, places particular reliance upon the fact that, at one stage, GONE and the Planning Inspectorate considered that the inquiry would have to be reopened. Letters of 18 April 2001, from Ms Burden of GONE to the Planning Inspectorate, and 2 May 2001, from Mr David Rose of the Planning Inspectorate to Ms Burden, acknowledged the appropriateness of reopening the inquiry. However, it is clear from Ms Burden's letter that it was not in the interests of fairness to the claimant that a reopened inquiry was seen to be appropriate. The letter said: G

H

> At the moment we do not know whether the Secretary of State is disposed to disagree with the Inspector's recommendations, and we cannot reach such a conclusion until the studies have been properly considered. Reference back is usually only undertaken when there is new evidence. As the retail study and audit were before the Inspector at the inquiry, and the local planning

> authority had been aware of them for some time, they do not appear to constitute new evidence. Although the appellant and Gateshead Metropolitan Borough Council did not dispute the findings of the new retail study and audit, other parties made written representations on the appeal, and there is no evidence that they have been afforded the opportunity to consider whether they accept the agreement on need and impact. In the light of this, the most appropriate course of action appears to be re-opening the public inquiry specifically to consider the issues of need and impact.

[30] Thus, it was purely to give parties other than the claimant and the council the opportunity to deal with the agreement on need and impact that GONE considered that a reopened inquiry was appropriate. The view that it took on this had nothing to do with any perception of fairness to the claimant. I can see nothing unfair in the procedure adopted by the Secretary of State. He was seeking additional conclusions on material that the claimant and the council had made available at the inquiry. There was no need for the claimant to be involved in that process at all. The opportunity that it was nevertheless given, to make representations on need and impact, and the fact that it was denied a sight of the inspector's report before doing so, did not create any unfairness, in my view. The claimant was clearly no worse off than if it had not been given such an opportunity. Mr Dove makes clear that he is not suggesting that the Secretary of State had attempted to influence the inspector in any way. Indeed, it is clear to me, having seen the correspondence and notes of discussion between GONE and the Planning Inspectorate, that the respective roles of the department and the Inspectorate were properly observed.

[31] It is clear, in my judgment, that the claimant has not been prejudiced in any way by the procedure adopted by the Secretary of State following the inquiry. The matter that the procedure was concerned with was the need for the development. Initially, the Secretary of State was not satisfied that the inspector had sufficiently investigated the evidence in accepting that the development was needed. At the end of the procedure, however, having received two supplementary reports from the inspector, the Secretary of State was able to say that he agreed with the inspector that retail need had been satisfactorily demonstrated. This conclusion was wholly in the claimant's favour: its case was accepted. I cannot see, in these circumstances, how it can be said that the claimant has suffered any prejudice. It is not contended that, in addressing retail need, the post-inquiry proceedings spilled over into other issues upon which the Secretary of State's conclusion were adverse to the claimant. In no respect have the claimant's interests been adversely affected by the procedure adopted by the Secretary of State. The requirement that a claimant must show substantial prejudice in a challenge based upon the failure to comply with a procedural requirement is contained in section 288(1) and (5) of the 1990 Act. In so far as the challenge is now made on wider grounds, the common law principle is the same. The court will not relieve a technical breach of natural justice: see *R* v *Chief Constable of Thames Valley, ex parte Cotton* [1990] IRLR 344. Here, there has, in the event, been no unfairness,

because the decision of the Secretary of State would not have been any A
more favourable to the claimant if he had dealt with the issue of need under some different procedure. The first ground of challenge accordingly fails.

Second area of challenge: Misinterpretation of policy

[32] Mr Dove's second ground of challenge asserts that the Secretary of B
State, in his decision, has misinterpreted his own policy on retail development. In para 19 of the decision letter, the Secretary of State said:

> 19. National policy on this issue is set out in PPG 6 and the Government's Response (May 2000) to the Second Report of the Environment, Transport and Regional Affairs Committee. Paragraph 1.12 of PPG 6 states that the Government expects that "developers and retailers will need to be more C
> flexible about the format, design and scale of the development, and the amount of car parking, tailoring these to fit the local circumstances". The Second Report emphasises that the Government has a clear policy that new retail investment should be used to strengthen existing centres and expects this paragraph of PPG 6 to be read in that light. PPG 6 is clear about the need for developers to demonstrate flexibility about the format of their D
> developments. The Government's policy is to favour the "class of goods", rather than the "format driven" approach to the application of the sequential approach, where the key question is whether there is any reason why such goods cannot be sold from town centres.

[33] Mr Dove says that PPG 6 does not put forward a class-of-goods approach for the purpose of the application of the sequential test. He also E
says that the government's response to the second report of the Environment, Transport and Regional Affairs Committee concerned food retail policy, and is irrelevant to comparison goods retailing. Accordingly, he says, the Secretary of State has erred in applying his policy on retail development as though it laid down the class-of-goods approach.

[34] I cannot accept this argument. I have quoted earlier the relevant F
paragraphs from PPG 6 and the government's response. While the second report of the select committee related to the environmental impact of supermarket development, it is, in my view, clear that the passage in the response that contrasts the format-driven and the class-of-goods approaches, and says that it is the latter that accords with PPG 6 is not to be read as confined to food retail policy. It is not said to be so confined. G
Paragraph 1.12 of PPG 6, to which express reference is made in the passage, is not confined to food retailing, and it seems to me inconceivable that the response is saying that this paragraph should be read as meaning one thing in the case of convenience goods and another thing in the case of comparison goods. Nor can I see anything inconsistent between what is said in the response and what is said in para 1.12 and other paragraphs of H
PPG 6. Paragraph 1.11 says that adopting a sequential approach means that the first preference should be for town-centre sites. Paragraph 1.12 says that developers and retailers will need to be more flexible about the format, design and scale of the development. The class-of-goods approach is to focus upon the range of goods proposed to be sold, and to consider

whether such goods could not be sold in a town-centre development or town-centre developments, if necessary of a format, design and scale different from the development proposed. I can see no inconsistency between this approach, which the Secretary of State evidently treats as being implicit in PPG 6, and the express words that are used. The second ground of challenge accordingly fails.

Third area of challenge: Application of the sequential test

[35] The third ground of challenge relates to the way in which the Secretary of State has applied the sequential test on the facts of the case. Mr Dove says that, in a number of respects, this application is perverse, or, if it is not perverse, is inadequately explained, and that certain material considerations have been left out of account by the Secretary of State.

[36] In her report, in paras 76 and 77, which I have quoted earlier, the inspector considered the sites that the council suggested were sequentially preferable to the appeal site. She noted that only three of the list 1 sites were suitable for retail warehouse development. She said that, since 80% of the trade draw came from within a 20-minute drive time from the appeal site, sites beyond this ceased to be contenders. She said also that, in her view, sites in local and district centres were not capable of fulfilling the same function as the appeal site because of consideration of sale. She went on:

> 78. The sequential approach must also be put into the context of the role and function of the MetroCentre which is recognised in the UDP and RPG as a centre of regional significance and as a preferred location for retail warehouse development. It is therefore an appropriate location for the identified need to be accommodated. The UDP was prepared in the light of the advice in PPG 6 para 1.10 requiring the identification of sites on a sequential basis. As need is not disputed, and in the absence of significant impact on any existing centre, then in my view this identified, allocated site in a regional centre must surely be an appropriate location for development of this nature, especially in view of its scale, nature and visual impact. (9).
>
> 79. In my opinion retailer investors would be likely to consider the appeal site differently, in terms of investment and likely footfall of customers, from a series of small scale schemes close to other town centres in the catchment area. This is perhaps why a number of the sites which are said to be available by the Council have remained undeveloped for many years. The Council's contention that the comparison goods spending available should be channelled into high street shops or small retail warehouse units spread throughout the area does not recognise the sector of the retail market which requires to trade from the retail warehouse format and which is acknowledged in PPG 6 and the UDP. (40,51)
>
> 80. There is no reason in theory why elements of the proposed scheme should not be provided on individual sites. This is a speculative development proposal with no identified end users. The Church Commissioners part of the appeal site, which already has planning permission, and the 3 sites on List 1 which are sites suitable for retail warehousing could accommodate much of the identified need for retail development. However, because of the history of long availability of the sites at Consett and Killingworth I agree with the Council that it cannot be

> assumed that these sites would be developed for retail development in the short term if the development on the appeal site were to proceed first. However, the availability of land for further retail development within the defined area of the MetroCentre is restricted. By the time the appeal proposals were built and occupied, the predicted demand for further development of this nature would be likely to be such as to result in the development of these other possible sites if they are in fact viable and sufficiently economically attractive. A B

[37] The inspector referred to the proposed Trinity Square redevelopment in Gateshead town centre. She said that it was smaller than the existing development, would be anchored by a major supermarket and would "meet a different type of comparison goods sector than the appeal proposals". She concluded: C

> 82. It seems to me that all the shopping centres in the region have had to adapt to the overwhelming presence of the MetroCentre over the years and it is clear that Gateshead Town Centre now functions primarily as a convenience goods centre and lower order comparison goods centre. The appeal proposals are not proposed as a new shopping destination, or as an addendum to a local or district centre but are destined for an allocated site within the defined retail area of the regional shopping centre to which Gateshead town centre has already adjusted. The MetroCentre already attracts large number of people and, as a result of the appeal proposals, the public transport and access by non-car modes would be improved. The appeal site is therefore, in my view, an acceptable and appropriate site for this form of retail development which would not compete directly with the Trinity Square scheme. (34,13,37) D E
>
> 83. I conclude that none of the other sites suggested by the Council are sequentially preferable to the appeal site which is identified for retail development within the defined boundary of the MetroCentre.

[38] The Secretary of State dealt with the question of sequentially preferable alternative sites in paras 19 to 29 of his decision letter. The first criticism that is made relates to the way in which the claimant's evidence on this was treated. Paragraph 20 says: F

> 20. The inquiry document "Retail Study – Hannington Works Site Gateshead" (CD102) was submitted by our client to GMBC in August 1999 as essential information in support of the proposals. Section 5 of the Study outlines how the sequential approach has been undertaken. Paragraph 5.3 states that only sites of approximately 3ha or above were examined as potential alternatives, on the grounds that one of the units comprising the development is likely to be around 9,290 sq m and would require a sizeable site, with the remainder of the scheme being divided into equal parts. G

[39] Then, at paras 23 to 25, the decision letter says: H

> 23. Paragraph 5.1 of CD102 states that the sequential search has been confined to locations within the 20 minute drive band of the site. However, paragraph 5.4 makes clear that not *all* the centres within this area have been included. Wallsend town centre, Killingworth town centre, Shields Road

> A District Centre in Newcastle and Prudhoe District Centre, for example, also lie within this area, but have been excluded from consideration without explanation.
>
> 24. Taking account of all this evidence, the Secretary of State considers that a more flexible approach with regard to format, design and scale of the development should have been adopted when carrying out the examination B of alternative sites. The analysis put forward by your client has unreasonably restricted the consideration of potentially suitable sites by imposing an artificially high minimum site size of 3ha, and by not including all the existing centres within the area of search.
>
> 25. Shortly before the inquiry took place, GMBC carried out a sequential analysis of alternative sites, on the basis of examination of all relevant development plans and emerging development plans, correspondence C with local authorities and site visits (GMBC5). The Secretary of State considers that this evidence is comprehensive and up to date. He has already set out his reservations on the evidence submitted by your clients (CD102) regarding the availability of sequentially preferable alternative sites in paragraphs 21-24 above. Accordingly, he is of the view that the Council's evidence on this matter should be given much greater weight than that of your client.

D

[40] There is no dispute that the primary evidence of the claimant on sequential alternatives was not the study submitted to the council in August 1999, but the supplementary evidence of the claimant's planning expert, Mr Michael Geoffrey Baker of GL Hearn. That evidence was prepared in January 2001, shortly before the inquiry. It included E Mr Baker's assessments of the alternative sites put forward by the council, including Killingworth Town Centre (B4), which was 2.71ha, Washington Town Centre (B5), 1.18ha, and Hermiston Retail Park (B6), 1.12ha. It dealt with other existing centres, including Shields Road District Centre (C4) in Newcastle. Thus, it addressed sites of less than 3ha and sites in existing centres, including Killingworth and Shields Road. I accept Mr Dove's F contention that, in expressing himself as he did in paras 23 to 25, the Secretary of State is revealing that he omitted from consideration this evidence of Mr Baker. Those paragraphs refer only to the 1999 study, and they give as the Secretary of State's reason for preferring the council's evidence on sequential alternatives the shortcomings in the 1999 study in failing to consider sites of less than 3ha and all sites in existing centres. It G seems to me inconceivable that the Secretary of State would have expressed himself as he did if he had paid regard to Mr Baker's supplementary evidence.

[41] Mr Timothy Mould submits that in those paragraphs (26 to 29) in which the Secretary of State expressed his conclusions on the suggested alternatives (and to which I will refer shortly), the Secretary of State must H be taken to have reached those conclusions in the light of all the material before him. I cannot accept this. By preceding his discussion of the alternative sites with the explanation in paras 23 to 25 as to why he preferred the council's evidence on these, the Secretary of State was, in my view, revealing his error in leaving out of account Mr Baker's supplementary evidence.

[42] The other criticisms that are advanced in relation to the Secretary of State's application of the sequential approach relate to his treatment of the particular alternatives that he considered in paras 26 to 29 of his decision letter. Mr Mould submits, and I accept, that these paragraphs are properly to be read in the light of the Secretary of State's overall conclusions in para 33. This paragraph follows on the Secretary of State's short expression of his acceptance of the need for the development:

> 32. The Secretary of State accepts that need for the development has been demonstrated, and is of the view that emerging RPG should be afforded little weight.
>
> 33. However, Government has a clear policy that existing city, town, district and local centres should be the focus for retail investment. The sequential approach to the selection of sites is of key importance in achieving this. In deciding this appeal therefore, the Secretary of State considers that the primary considerations are the extent to which the sequential approach has been satisfactorily carried out, and the availability of sequentially preferable suitable alternative sites. For the reasons given in paragraphs 21-29 above, he is of the view that your client has not demonstrated sufficient flexibility about the format of the development. There is strong evidence that suitable alternative sequentially preferable sites are available. In addition, it is likely that some of the retailing your client proposes will not be bulky goods, and could be carried out from town centres. The Secretary of State therefore considers that the failure to carry out the sequential approach satisfactorily, and to fully consider alternative sequentially preferable sites, is a material consideration of great importance, sufficient to override the proposal's consistency with the development plan, and all the other factors in its favour.

[43] Paragraphs 26 to 29 say:

> 26. He [that is the Secretary of State] notes (GMBC5, GMBC7, GMBC13, GMBC18) that there are sites with planning permission for retail warehouses in Killingworth Town Centre (B4), on the edge of Consett town centre at Hermiston Retail Park (B6), and in the Shields Road District Centre (C4) in Newcastle. Both Consett and Killingworth are defined as town centres, and there is no evidence that retail warehouse development of the size already permitted on B4 and B6 would be out of scale with these centres. Shields Road is a larger District Centre which is the focus of a significant regeneration scheme, where planning permission for upwards of 10,000 sq m overall of retail development of different types has recently been granted.
>
> 27. The Secretary of State has already concluded in paragraph 22 of this decision letter that some of the goods which could be sold from the proposed development are commonly sold in town centres. Information was before the inquiry (IR50) that planning permission was granted in November 2000 for redevelopment in Gateshead Town Centre to provide 26,100 sq m of retail floorspace. There is therefore no reason why some of the retail development proposed by your clients could not be carried out from premises in Gateshead Town Centre.
>
> 28. In addition, the Secretary of State notes that there are other sequentially preferable sites potentially available in the longer term. The Sunderland Road site in Gateshead (C3), for example, which is capable of functioning as an edge of town centre site subject to new pedestrian and bus

> links, has recently been the subject of developer interest for 11,684 sq m non food DIY warehouse development. There is 1.18 ha of land available within the nearby Washington Town Centre (B5) which is allocated in the Sunderland UDP for retail and commercial development.
>
> 29. Sites B4, B6 and C4 could provide for some 11,531 sq m of retail warehousing, are suitable for this use and could be made available in the short term. Some of the retailing proposed could be located in Gateshead Town Centre. In the light of this, the Secretary of State is of the view that there are sequentially preferable suitable sites available within the 20 minute drive time area around the application site which could easily accommodate at least half of the retail development proposed. Other land is potentially available in the longer term capable of providing for the rest of the floorspace proposed. The key issue is whether sites such as C3 and B5 are likely to become available within a reasonable period of time. Evidence before the inquiry (GMBC8), which formed part of the basis for the agreement between the parties on the issue of need, concluded that a small positive expenditure capacity would only arise by 2004, and comfortable expenditure capacity by 2006, to support all of the currently identified commitments, including the turnover requirement of the present application. In the light of this, the Secretary of State considers that there is no compelling need to provide all of the proposed floorspace immediately, and that the period up to 2006, that is the next five years, represents a reasonable period of time for suitable alternative sites to be brought forward. There is no evidence that C3 and B5 could not be made available for development within this timescale.

[44] Mr Dove submits that, in concluding that Killingworth (B4), Hermiston (B6) and Shields Road (C4) were suitable for the proposed use, the Secretary of State had failed to explain why he was differing from the inspector, who had rejected them as unsuitable. Mr Mould submits that the Secretary of State gave adequate reasons in this respect. The Secretary of State, he says, disagreed with the inspector on the basis of policy. The inspector essentially adopted the format-driven approach, while the Secretary of State applied the class-of-goods approach, and concluded accordingly that the sites were suitable in terms of the sequential approach.

[45] I do not think that the difference between the inspector and the Secretary of State is to be explained as simply as Mr Mould suggests. I am not sure that the inspector was in fact adopting the format-driven as distinct from the class-of-goods approach. She said (para 77) that South Shields (B2), Whitley Bay (B3), Hermiston (B6) and Shields Road (C4) ceased to be contenders because of their location. She said also that sites in district and local centres were not capable of fulfilling the same functions as the appeal proposals because of the need for new development to be in scale with the centre. She was of the view (para 78) that the appeal site, as an allocated site in a regional centre, was certainly suitable for the development proposed. She said (para 80) that there was no reason why elements of the proposed scheme should not be provided on individual sites, but she thought (para 79) that retail investors would be likely to consider the site differently from a series of small-scale schemes close to other town centres in the catchment area, and that that was perhaps why

a number of the sites said to be available by the council had remained undeveloped for many years. She thought that the council's contention that the comparison good spending available should be channelled into high street shops or small retail warehouse units, spread throughout the area, did not recognise the retail sector that required to trade from the retail warehouse format, and which was acknowledged in policy. A

[46] The inspector was clearly attaching weight to what she regarded as a requirement for a sector of the retail market to trade from retail warehouses. It was undoubtedly material for her to consider the preferences of retailers in this respect. Paragraph 1.12 of PPG 6 says that local planning authorities should be sensitive to the needs of retailers and para 3.3 states: B

C

> 3.3 Some types of retailing, such as large stores selling bulky goods, may not be able to find suitable sites either in or on the edge of town centres. In such cases, the local planning authority should still seek to ensure that such developments are located where they will be easily accessible by a choice of means of transport. (See also para 1.17.)

[47] The inspector did not, however, rule out the council's suggested alternatives solely because they could not provide for large-format retail warehouses. She recognised that elements of the proposed development could be accommodated on individual sites, but it appeared to her that the alternatives were unattractive to retail investors. Furthermore, those sites that were on the edge of, or outside, the 20-minute drive zone were not, as she put it, contenders. D

E

[48] In the light of these conclusions on the part of the inspector, the reasons given in paras 26 to 29 by the Secretary of State for the different conclusion that he came to were, in my judgment, deficient in three respects. First, he did not say what weight, if any, he attached to the demand for large retail warehouses on the part of a sector of the retail market. Adopting the class-of-goods approach to the sequential test clearly does not render immaterial this particular consideration, which is expressly recognised in para 3.3 of PPG 6. Paragraph 1.12 requires local planning authorities to be sensitive to the needs of retailers. In view of the weight attached to this consideration by the inspector, it was necessary, in my view, for the Secretary of State to explain how the judgment that he had formed took account of it. F

G

[49] The second deficiency in the Secretary of State's reasoning, in my judgment, is that he failed to explain his views on the suggested unattractiveness of alternative sites to retail investors. In view of the conclusions of the inspector and the express reference to viability in para 1.12 of PPG 6, it was not sufficient for him, in my view, to deal with alternative sites solely in terms of their availability. The claimant was entitled to be told whether the Secretary of State took a different view from the inspector on the question of the viability of alternative sites. H

[50] The third deficiency, in my judgment, is that the Secretary of State failed to say why he rejected – as, impliedly, he must have rejected – the inspector's conclusion that certain of the alternatives ceased to be

contenders on account of their location. The evidence to which the inspector had regard was based upon what I understand to be a conventional study of retail capacity and impact. Such a study looks at the trade that would be drawn to the proposed development from a catchment area centred on the appeal site. The catchment area is defined in terms of drive-time bands, and the trade draw relates to the expenditure of the population within the catchment area on the range of goods under consideration. It follows that a catchment area, and consequently the population within it, the expenditure and the trade draw are all site-specific. Each site will have its own catchment area, and, in general, the further the two sites are apart, the less their catchment areas will overlap. It was this point that the inspector evidently had in mind when saying, as she did in para 77, that, since 80% of the trade draw came from within the 20-minute drive band, four list 1 sites, and most of the list 2 sites, ceased to be contenders. In his decision letter, the Secretary of State, at para 29, identified three particular sites as being suitable and available. Of these, Hermiston (B6) and Shields Road (C4) were among the sites that the inspector had discounted because of their location. In disagreeing with the inspector on the point, the Secretary of State needed, in my view, to explain why these sites were to be regarded as suitable alternatives despite their distance from the appeal site, and, consequently, the different catchments that they would serve.

[51] A further deficiency in the reasoning of the Secretary of State, Mr Dove submits, is to be found in para 29, where the conclusion was expressed that there was no compelling need to provide all of the floorspace immediately, and that the period up to 2006 represented a reasonable period of time for suitable alternative sites to be brought forward. This conclusion was there expressed to have been reached in the light of the following:

> Evidence before the inquiry (GMBC8), which formed part of the basis for the agreement between the parties on the issue of need, concluded that a small positive expenditure capacity would only arise by 2004, and comfortable expenditure capacity by 2006, to support all of the currently identified commitments, including the turnover requirement of the present application.

[52] Mr Dove points out that the claimant's case on need distanced itself from the council's analysis, which was derived from the April 1999 study. This is clear from the way in which the inspector recorded the claimant's case at para 36, which I have quoted earlier. The GL Hearn study of November 2000 was said to show that there would be sufficient capacity for the appeal scheme and other commitments by 2003, with a surplus beyond this of £123.10m of retail expenditure. Indeed, sufficient capacity to support the appeal scheme was said to exist by the end of 2001. In her second supplementary report (at para 6, which I have also quoted earlier), the inspector said that the worst case postulated was that, by 2006, expenditure growth would comfortably support all the identified commitments, while a less cautious calculation suggested that, by 2003, there would be a residual of £36.7m, even when confirming the

analysis to the 20-minute catchment area. In her first supplementary report, the inspector had said in para 6: A

> The parties agreed (Doc GMBC8), on the basis of the Retail Study, that an objection to the appeal proposals based on the failure to demonstrate need could not be sustained and would not be pursued.

[53] That, manifestly, was not a statement that the claimant and the council were agreed that capacity would not arise until 2004, with comfortable expenditure capacity by 2006. The other passages that I have referred to in the inspector's report and her second supplementary report made clear that the claimant was taking a more optimistic view, and that the inspector was not rejecting this view. B

[54] In these circumstances, there was, in my judgment, a need for the Secretary of State to explain why he based himself upon the council's view on the date when capacity would arise, rather than upon the claimant's view, derived from its study of November 2000. His failure to do so is a further defect in his reasoning. C

[55] In para 27 of his decision letter, the Secretary of State said that there was no reason why some of the retail development proposed could not be carried out in the 26,100m^2 of retail space proposed in the Gateshead Town Centre redevelopment. Mr Dove pointed out that the redevelopment was to replace existing floorspace, and the resulting non-food area would in fact be less than at present. I agree that the Secretary of State's reliance upon the redevelopment is unexplained and, on the face of it, illogical. Paragraph 29 makes plain that he was viewing the redevelopment floorspace as available to meet capacity requirements (and not simply to satisfy the requirement of retailers for newly developed floorspace). His unexplained reliance upon the Gateshead redevelopment is a further defect in his reasoning, in my view. D E

[56] The final element of Mr Dove's third ground of challenge derives from the fact that, as he points out, no reference at all is made in the decision letter to the appeal decision of 23 February 2000 on the Church Commissioners site. That appeal related to a substantial part of the appeal site. The policy issues were the same, and the need issues were also the same, albeit related to a rather smaller development. The Secretary of State decided, in that case, that planning permission should be granted. Mr Dove's first submission is that, since the Secretary of State did not refer to the Church Commissioners appeal decision, it is to be inferred that he left it out of account. I find it impossible to draw such an inference. It seems to me to be inconceivable that the Secretary of State could simply have overlooked such an obviously relevant recent decision, and it would only be right to infer that he had deliberately left it out of account if there was some inescapable inconsistency between the grant of permission in the one case, and its refusal in the other. It does not seem to me, however, that one can say more than that there is an apparent inconsistency between the two decisions. But it is because of this apparent inconsistency that I accept the second way in which Mr Dove puts the challenge – that it was incumbent upon the Secretary of State to explain what it was that F G H

distinguished the earlier proposal from the present one, so as to justify permission in the one case, and to require refusal in the other.

[57] Mr Dove relies upon *North Wiltshire District Council* v *Secretary of State for the Environment* [1992] 3 PLR 113, where, at pp122F-123B, Mann LJ said:

> In this case the asserted material consideration is a previous appeal decision. It was not disputed in argument that a previous appeal decision is capable of being a material consideration. The proposition is in my judgment indisputable. One important reason why previous decisions are capable of being material is that like cases should be decided in a like manner so that there is consistency in the appellate process. Consistency is self-evidently important to both developers and development control authorities. But it is also important for the purpose of securing public confidence in the operation of the development control system. I do not suggest, and it would be wrong to do so, that like cases *must* be decided alike. An inspector must always exercise his own judgment. He is therefore free upon consideration to disagree with the judgment of another but before doing so he ought to have regard to the importance of consistency and to give his reasons for departure from the previous decision.
>
> To state that like cases should be decided alike presupposes that the earlier case is alike and is not distinguishable in some relevant respect. If it is distinguishable then it usually will lack materiality by reference to consistency although it may be material in some other way. Where it is indistinguishable then ordinarily it must be a material consideration. A practical test for the inspector is to ask himself whether, if I decide this case in a particular way, am I necessarily agreeing or disagreeing with some critical aspect of the decision in the previous case? The areas for possible agreement or disagreement cannot be defined but they would include interpretation of policies, aesthetic judgments and assessment of need. Where there is disagreement then the inspector must weigh the previous decision and give his reasons for departure from it. These can on occasion be short, for example, in the case of disagreement on aesthetics. On other occasions they may have to be elaborate.

[58] Mr Mould submits that the only question is whether it could be ascertained from the decision letter why it was that the Secretary of State was refusing planning permission, despite having granted it in the Church Commissioners appeal. If it could be so ascertained, it did not matter that the earlier decision was not referred to. In my judgment, the need for an express explanation of an apparent inconsistency between the decision under consideration and an earlier decision will depend upon the circumstances. If the explanation for the inconsistency is obvious, a formal statement of it will be unnecessary. Where the inconsistency is stark and fundamental, as it seems to me it is in the present case, it will, in my judgment, usually be insufficient to leave it to the reader to infer the explanation for the inconsistent decisions. The reason for this is that unless the decision maker deals expressly with the earlier decision and gives reasons that are directed at explaining the apparent inconsistency, there is likely to be a doubt as to whether he has truly taken the earlier decision into account. In the present case, moreover, the inspector had thought it

appropriate to place reliance upon the earlier decision, and had referred to A
it no less than 33 times in the course of her report. The claimant was entitled, in these circumstances, to an express explanation on the Secretary of State's part, and it has been substantially prejudiced by the lack of such an explanation.

[59] In my judgment, the reasons given by the Secretary of State for disagreeing with the inspector and dismissing the appeal are, in a number B
of respects, seriously inadequate, and there is no doubt that the claimant has been prejudiced as a result. In addition, the decision letter shows that the Secretary of State left out of account the claimant's supplementary evidence on the sequential test, which pointed to a different result from the one that he reached on the basis of the council's evidence. The application is granted.

Claim allowed.

Edinburgh City Council *v* Scottish Ministers

COURT OF SESSION: OUTER HOUSE
Lord Eassie
17 September 2002

Listed building enforcement notice – Planning (Listed Buildings and Conservation Areas) (Scotland) Act 1997 – Planning consent granted subject to conditions – Whether reporter erred in law – Whether conditions invalid and unenforceable

The appellant planning authority issued an enforcement notice on the owner of a B listed, two-storey terraced house situated within a conservation area, alleging a breach of planning control in respect of a ground-floor front room that was used as an office and guest bedroom during the tourist season. The notice stated that the erection of full-height partition walls forming an en suite bathroom, without listed building consent, was contrary to section 6 of the Planning (Listed Buildings and Conservation Areas)(Scotland) Act 1997, and set out the remedial steps requiring the removal of the walls. The owner appealed against the enforcement notice to the Scottish Ministers, under section 35 of the 1997 Act. The reporter hearing the appeal concluded that the unauthorised works affected the character of the building as a building of special architectural or historical interest, and required listed building consent. He concluded that consent should be granted, but attached a condition requiring the removal of the partition walls within one month of whichever of the following events that first occurs: (i) the permanent cessation of use of the room as a bedroom; (ii) removal of the net curtains; or (iii) removal of the privet hedge on a footpath outside the room. The planning authority appealed against that decision, contending that: (i) the reporter had erred in law, and had misdirected himself, by failing to have regard to the presumption against development affecting the character of a listed building, and by having regard to irrelevent considerations; and (ii) the conditions imposed were uncertain, unenforceable, unreasonable or irrelevant to listed building control.

Held The appeal was allowed.

The reporter misdirected himself in law. The conditions were flawed and unenforceable. It would be difficult to know whether the use of a room within domestic premises as a bedroom had ceased "permanently". The second subcondition (net curtains) was also unsatisfactory. It raised the issue of whether an alteration forming part of the heritable or immovable property may properly be given consent subject to conditions in relation to the use of soft furnishings such as curtains, since planning conditions and listed building conditions relate essentially to immovable property. The third subcondition (hedge), while a less ephemeral feature of the property, also suffered problems in its drafting: see [20] and [21]. It is important that those granting planning consent subject to conditions give care to the draftmanship involved in framing such conditions. The underlying problem was that, while purporting to decide whether listed building consent for the works in issue should be granted, the reporter in fact addressed a different question, namely whether it was reasonable that the works, having been erected without a prior grant of listed building consent, should now be removed: see [24]. In some cases, the distinction between asking oneself whether it was reasonable that an offending structure be removed, or whether a proposal *ab ante* to construct such a structure should receive consent, may be a fine one. But the two questions are truly different. To apply the former test may risk deploying, in favour of the property owner who has carried out the development or alteration without the consent required by law, a practical advantage not enjoyed by those owners who are more careful and scrupulous in their attention to the requirements of the legislation. It also invites the imposition of inappropriate conditions directed, not to the grant of consent, but to the cessation and removal of an unauthorised alteration to a listed building for which consent might never have been granted *ab ante*. To apply the former test is an error of law: see [25].

Cases referred to in the judgment

British Airports Authority v *Secretary of State for Scotland* 1979 SC 200; [1980] JPL 260

Edinburgh City Council v *Secretary of State for Scotland* [1997] 1 WLR 1447; [1998] 1 All ER 174; [1997] 3 PLR 71; [1998] JPL 224; 1998 SC (HL) 33, HL

Freeport Leisure plc v *West Lothian Council* 1999 SLT 452

Simpson v *Edinburgh Corporation* 1960 SC 313

South Lakeland District Council v *Secretary of State for the Environment* [1991] 1 WLR 1322; [1992] 1 All ER 45; (1991) 89 LGR 857; (1991) 62 P&CR 617; [1991] 2 PLR 97; [1991] JPL 654, CA

Appeal under sections 57 and 58 of the Planning (Listed Buildings and Conservation Areas) (Scotland) Act 1997

This was an appeal under sections 57 and 58 of the Planning (Listed Buildings and Conservation Areas) (Scotland) Act 1997 by Edinburgh City Council, against a decision of the Scottish Ministers determining an appeal, under section 35 of the Act, against a listed building enforcement notice issued by them on the owner of a building.

R Douglas Armstrong (instructed by the solicitor to Edinburgh City Council) appeared for Edinburgh City Council.

Ruth Crawford (instructed by the solicitor to the Scottish Executive) represented the Scottish Ministers.

The following judgment was delivered.

LORD EASSIE:

[1] This is an appeal presented pursuant to sections 57 and 58 of the Planning (Listed Buildings and Conservation Areas) (Scotland) Act 1997 (the Act) against a decision of a reporter appointed by the Scottish Ministers to hear and determine an appeal to the Scottish Ministers, under

section 35 of the Act, against a listed building enforcement notice issued by the local planning authority. The appeal to the Court of Session is brought by the planning authority, and has been remitted for hearing in the Outer House in terms of Rule of Court 41.44(1).

[2] The listed building with which these proceedings are concerned is situated in Edinburgh, and is described by the reporter, in para 2 of his decision letter, in these terms:

> The appeal property is part of a "B" listed classical 2-storey terrace of original townhouses on the north east side of Bellvue close to its junction with East Claremont Street within the New Town Outstanding Conservation Area. At the time of my inspection the originally square shaped room in question was in use as an office, but I understand that in the tourist season it is used as a bedroom as part of a guest house. It has an elaborate plaster ceiling and extended cornice. In one corner an area has been partitioned off to form an en-suite bathroom; its partition walls do not extend as far up as the cornice. The room has windows on the principal elevation facing south west: at the time of my inspection, these windows had net curtains. The front garden is about 5m deep, and there is a low stone wall and a high privet hedge at the heel of the footpath.

[3] Put shortly, section 34 of the Act empowers a planning authority to serve a listed building enforcement notice where there has been a contravention of section 8(1) or (2) of the Act. In turn, section 8(1) of the Act refers to a contravention of section 6 of the Act, which states that:

> Subject to the following provisions of this Act, no person shall execute or cause to be executed any works for the demolition of a listed building or for its alteration or extension in any manner which would affect its character as a building of special architectural or historic interest, unless the works are authorised.

The contravention of section 6 alleged in the present case was labelled in the listed building enforcement notice as:

> Without listed building consent, the erection of full height partition walls to form an en-suite bathroom within the front, right hand (as viewed from the street) principal room on the ground floor of the said townhouse...

The steps required to remedy the breach of listing building control were specified in the enforcement notice as follows:

> Remove the unauthorised partition walls forming the en-suite bathroom from the front, right hand (as viewed from the street) principal room on the ground floor of the townhouse at [the address in question] and repair any damage caused to the architectural features of the room by the works to erect or remove the unauthorised partition walls.

[4] In appealing to the Scottish Ministers against the listed building enforcement notice the owner invoked three of the possible grounds of appeal set out in section 35 of the Act, namely grounds (b), (c) and (e) which are in these terms:

(b) that the matters alleged to constitute a contravention of section 8(1) or (2) have not occurred; A

(c) that those matters (if they occurred) do not constitute such a contravention;

...

(e) that listed building consent ought to be granted for the works, or that any relevant condition of such consent which has been granted ought to be discharged, or different conditions substituted; B

[5] The point taken by the owner under ground (b) was to the effect that the description of the contravention wrongly described the partitions as being "full height", whereas the structure of which complaint was made did not extend fully to the ceiling. The reporter dismissed this ground of the appeal to him on the basis that this misdescription of the offending C structure did not create any doubt as to the structure in question, and the notice could properly be corrected. In the present proceedings, no issue arises respecting the reporter's decision on that point.

[6] The second ground – para (c) of section 35(1) of the Act – raises the question of whether the allegedly offending works, namely the creation of a toilet and bathroom within the front room of the ground floor of the D house, in a partitioned enclosure not extending to full height, constituted something requiring listed building consent, as being an alteration that "in any manner would affect... [the] character [of the listed building] as a building of special architectural or historical interest". The reporter concluded that the offending works did indeed have such an effect upon the character of the building and, accordingly, did require listed building E consent. He expresses his conclusion, in para 19 of the decision letter, thus:

> In my opinion, the character of the appeal property is made up of a number of elements, including the proportions of the principal rooms facing the front and the moulded ceiling and extended cornice of the room in question. The ceiling and cornice are unaffected, but my inspection confirmed that the F room's proportions have indeed been altered by the erection of the en-suite partition walls. Had the room been at the back away from the street frontage and from public view, it might be argued that the building's overall character had not been affected. However, it is undoubtedly a "principal room" whose windows are on the front elevation, and I conclude that the appeal property has been altered in a manner that has affected its character as a listed building and that listed building consent is therefore required. The appeal G under this ground also fails.

It may inferred from the reporter's placing of the words "principal room" in quotation marks that he intended a reference to a passage from *Historic Scotland's Memorandum of Guidance on Listed Buildings and Conservation Areas*, from which the reporter had previously quoted, in para 6 of his H decision letter, in these terms:

> The original plan form should always be respected and all major works of alteration limited to areas of secondary importance. It is consequently seldom possible to subdivide into several smaller units in an acceptable manner a larger property which retains its original plan form... The entrance

> hall, main stair and principal apartments, especially where these are publicly accessible or lie to the front of the building and may be seen through windows from the street outside, should never be sub-divided or opened up. Where this has happened reinstatement should always be encouraged… Where new internal walls cannot be avoided, they should not cut through timber detailing or enriched plaster decoration, but should be scribed around them to facilitate reinstatement at a later date…

[7] Having thus concluded that the structure did indeed affect the character of the appeal property as a building of special architectural or historic interest, the reporter then addressed himself to the issue as to whether it should receive listed building consent. He concluded that the offending structure should receive such consent, but subject to a condition. The condition, which comprised subheads, was that:

> the partition walls shall be removed and the room restored to its original proportions within one month or whichever of these matters first occurs:
> (1) the permanent cessation of the use of the room as a bedroom;
> (2) the removal of the net curtains from the windows facing Bellvue;
> (3) the removal of the privet hedge at the heel of the footpath.

[8] The present appellants, the planning authority, challenged the validity of this branch of the reporter's decision. Before summarising the submissions for the planning authority, it is convenient to set out the reporter's reasons for reaching the view that listed building consent should be granted on those conditions. In para 20 of the decision letter, the reporter defined the determining issues on this branch of the appeal to him as being:

> whether the alterations that have been carried out have an adverse effect on the character of the listed building; and, if so, whether there are other considerations which might nonetheless justify approval of the works.

The reporter then went on to say, in paras 21 to 23:

> 21. The appeal property is an important listed building in its own right, and as a significant part of both a classical listed terrace and an outstanding conservation area. The room in question is a "principal room" of classical proportions, which have been affected by the creation of the en-suite bathroom. However, while the integrity of the appeal property as a listed building has been compromised, I am not convinced that this is so great as to justify the immediate removal of the offending structure. You [the owner] have been careful to ensure that the elaborate ceiling and extended cornice have not been damaged. I am also aware that, while the plumbing and internal fittings were presumably intended to be permanent, the partition walls are reversible and capable of being removed without damage to the ceiling or cornice, and you have suggested the possibility of a temporary permission. Crucially, in my opinion, the tall hedge at the heel of the footpath and the net curtains combine to make the partition walls virtually invisible from outside.
>
> 22. I accept that what you have done is inconsistent with the guidance and policies which I have set out in paras 5-8 above, but I consider that the lack of damage to the ceiling and cornice combined with the almost complete

> invisibility of the partition walls from the front allow an exception to be A made in this case without the creation of an undesirable precedent. I believe that the limited nature and impact of the works – together with the facts that the property is not in category A and that the room at times provides a service for visitors to the city – justify a pragmatic, rather than a purist approach. Your appeal under this ground therefore succeeds.
>
> 23. However, I do not believe that an unrestricted consent is appropriate. B Circumstances may change whereby the room is no longer required to be used as a bedroom, or the net curtains and/or the privet hedge are removed, resulting in the impact of the partition walls on the proportions of the room and hence also on the character of the listed building, terrace and conservation area becoming significantly more noticeable. For this reason, I consider your suggestion of a temporary permission to be relevant. I am aware of the guidance on temporary permissions in SODD Circular 4/1998, C which states: "... it is undesirable to impose a condition requiring the demolition after a stated period of a building that is clearly intended to be permanent", but I believe that the circumstances of this case justify such a condition. You have suggested a period of 5 years, but I consider it more appropriate to link removal of the en-suite structure to the cessation of the room's use as a bedroom, and the removal of the net curtains and the hedge.

D

[9] Counsel for the planning authority advanced two principal submissions. First, the reporter erred in law, and misdirected himself, by failing to have regard to the presumption against development affecting the character of a listed building, and by having regard to other, irrelevant considerations. Second, the conditions imposed were individually and collectively uncertain, unenforceable, unreasonable or irrelevant to listed E building control, and they could not individually or collectively be severed from the grant of listed building consent.

[10] In advancing his first principal submission, counsel for the planning authority referred to the Scottish Office published guidance in paras 42 and 43 of NPPG 18 (No 6/5 of process). He pointed to that part of those paragraphs that indicated the primary, statutory requirement (in F section 14(2) of the Act) to have special regard to "the desirability of preserving the building or its setting or any features of special architectural or historic interest which it possesses". Counsel then stressed the guidance in para 43, that there should be "a presumption against development that adversely affects the character of a listed building or its setting". The reporter, he said, made no reference to that presumption, and G did not relate it to the facts that he had found, namely that: the building was an important listed building; the room in which the structure had been erected was a principal room; the structure had affected the proportions and dimensions of that room; and its integrity as a listed building had been compromised. Given those findings, the reporter failed to explain how the presumption against development affecting the H character of the listed building was overcome, other than by reference to what the reporter described as "crucial", namely the current existence of a hedge and the then current use of net curtains in the windows of the principal room in issue. If, as the reporter had plainly found, the structure adversely affected the character of the building, it was irrelevant to say

that the provision of curtains might make it not readily visible to the passing public on the footpath. The implication of the reporter's approach was that consent should be granted for any interior alteration, however seriously affecting the character of a listed building, provided that observation of its adverse effects were screened from the passing public by means of curtains. The reference to the appeal property's not being a category A listed building was also an irrelevant consideration, since the category was not a guide to the building's sensitivity, and, in that respect, counsel referred to in *Listed Buildings* by Suddards & Hargreaves (3rd ed) para 4.11. Further, in proceeding initially upon the view that the hedge and curtains provided a degree of invisibility to the passing public, the reporter thereafter proceeded upon a basis inconsistent with that view, and the memorandum of guidance, in his contemplation of the provision of a service to visitors by the use of the room as a bedroom during the tourist season. The room would therefore be, to that extent, publicly accessible and inconsistent with his earlier erroneous criterion of external visibility.

[11] Counsel for the planning authority then turned to the terms of the condition imposed by the reporter, and particularly the three subheads by which the reporter made provision for bringing the consent to an end.

[12] As respects the first subhead ("permanent cessation of the use of the room as a bedroom"), counsel for the planning authority made the preliminary observation that, at the time of the reporter's inspection, the remainder of the principal room in question was not in use as a bedroom but as an office. The occupier apparently intended its use as a bedroom only on some seasonal basis. How, asked counsel, is a local planning authority to know whether a room within domestic premises is being used as a bedroom? Did use as a bedroom exclude use for other purposes, such as a study? How could it be determined that a cessation of use as a bedroom was "permanent"? There was no obligation on a proprietor to seek planning consent for change of use for rooms within domestic premises. These matters individually and collectively rendered the condition unenforceable in any reasonable sense. Further, asked counsel, how could the use of the room and its partitioned subdivision be material to the deleterious effect of the alterations upon the character of the listed building? This meant that the condition was not only unenforceable but also irrelevant.

[13] As respects the second subhead of the condition (net curtains), counsel for the planning authority submitted that this was similarly flawed as unreasonable, unenforceable and irrelevant. As a matter of drafting, the condition said nothing about whether the net curtains were required to be kept closed. The curtains would be removed for cleaning, and the question would then arise as to whether the consent must thereby come to an end. More importantly, the resort to the use of soft furnishings as a planning condition was wrong in principle, if only by reason of the difficulties inherent in enforcement and policing. Removal of net curtains was again not something for which any application for consent would be required.

[14] Adverting to the third subhead, counsel submitted that, as with the net curtains, this was also irrelevant, both to the question of whether the character of the building had been adversely affected and to the extent of that adverse effect. Removal of the privet hedge was likewise something for which no application for consent would be required. Further, the condition said nothing about reducing the height of the hedge, or, indeed, its demise from natural causes. A B

[15] Counsel went on to state that the granting of conditional planning consents was the subject of Circular 4/1998 (No 6/6 of process), which set out appropriate tests. Those tests were also applicable to the grant of listed building consent – a proposition with which counsel for the respondents intimated her agreement. Reference was also made to para 44 of NPPG 18, which states that conditions: C

> must be necessary, relevant to preserving the building or its setting or any features of special architectural or historic interest, enforceable, precise and reasonable in all other respects.

That paragraph contained a brief summary of the import of Circular 4/1998, but counsel referred in particular to para 105 of the latter, in relation to the principles to temporary permissions. Counsel further submitted that the condition, with its subheads, was not properly severable from the reporter's decision to grant listed building consent, and, in regard to the general issue of the severability of planning conditions, counsel referred to *British Airports Authority* v *Secretary of State for Scotland* 1979 SC 200. D E

[16] In opposing the motion by counsel for the planning authority that the reporter's decision be quashed, counsel for the respondents submitted that the decision was reasonable, correct in law, and, therefore, *intra vires*. As part of a relatively detailed examination of the statutory provisions of the Act, she stressed that the facts that a building was a listed building, and that the works proposed affected its character as a building of special architectural or historical interest, did not mean that the proposed works, so far as affecting the character of the building, could not be permitted. In deciding whether to grant listed building consent, special regard had to be paid to the desirability of preserving, in this case, "any features of special architectural or historical interest" that the building possessed: see section 14(2) of the Act. The verb "to preserve" simply meant "to keep safe". Reference was made in that connection to *South Lakeland District Council* v *Secretary of State for the Environment* [1991] 1 WLR 1322[1]. The requirement "to have regard" to a plan or requirement did not equiparate with slavish adherence, and, in that respect, counsel referred to *Simpson* v *Edinburgh Corporation* 1960 SC 313. It would be perfectly proper for a reporter to assess all the relevant materials and circumstances and to exercise judgment as to whether to grant consent: cf *Edinburgh City Council* F G H

[1] [1991] 2 PLR 97

A v *Secretary of State for Scotland* 1998 SC (HL) 33[1]. While material such as NPPG 18 and the memorandum of guidance might be of help to a reporter, they were not legally binding: see *Freeport Leisure plc* v *West Lothian Council* 1999 SLT 452. What one was required to do was to ascertain whether the reporter had failed to identify the right test, and, if correctly identified, whether its application was erroneous or unreasonable.

B [17] Counsel for the respondents then turned to the decision letter and the reporter's reasons for concluding that grant of listed building consent, subject to the condition, might be justified. She pointed out that, in para 19, ,the reporter had noted that, had the room been at the back of the house, away from the frontage and the public view, it might have been contended that, not being a principal room, the building's overall character had not C been affected by the offending structure. Accordingly, she submitted, visibility from the street was a relevant factor in a decision whether to grant listed building consent. Since, as the reporter noted, the ceiling and cornices were not damaged, the partitions forming the walls of the bathroom were reversible (by which one assumes "removable"), and the disputed structure, by reason of the existence of the net curtains and the D hedge was not readily visible from the street, the reporter was entitled to conclude that the works were limited in nature and impact. The works being limited in those two respects, it could not be said that he had applied the wrong tests with the result that his judgment was flawed on that account.

[18] As regards the condition attached, counsel for the respondents E submitted that the condition, including all its subheads, was imposed only in the event of future change of circumstances. One of the elements prominent in the reporter's decision had been visibility from the street. Subheads (2) and (3) were pertinent to visibility and therefore relevant. Subhead (1) reflected the fact that the bathroom and toilet that had been installed in the principal room were so installed because of an intention to F use the rest of that room as a bedroom, and the need for the former would cease upon a decision not to use the latter as a bedroom. Counsel further submitted that, on a reasonable view, all the subheads of the condition were capable of enforcement. Whether a room was being used as a bedroom was a question of fact that might be ascertained. Whether the privet hedge had been removed was readily ascertainable. Whether the G net curtains had been removed was also readily ascertainable by means of outside inspection, and it was implicit in the decision that a temporary removal of the net curtains for the purposes of cleaning would be permissible.

[19] In addressing these competent submissions, I find it convenient first to examine the terms of the condition that the reporter attached to the H grant of listed building consent and that he evidently regarded as highly material to his decision to grant consent. Subject to any issue of the

[1] [1997] 3 PLR 71

separability of that condition or its subheads, it appears to me that the A propriety of the decision to grant listed building consent may be viewed as turning, in the first place, upon the validity of that condition, with all its subheads. The reporter clearly did not intend to grant any temporally unrestricted consent, and regarded three particular matters as being each, in themselves, decisive as respects the duration of the consent.

[20] Despite what was said by counsel for the respondents in her B defence of the reporter's decision, it is, in my view, evident that the terms of the condition that he sought to impose present difficulties. The first subhead is couched in terms of the permanent cessation of the use of the room – by which I assume the remainder of the principal room subdivided by the structure – as a bedroom. There is, of course, no planning control on the use to which a householder or the owner of a guest house may put the C individual rooms within a house, and, therefore, this consent immediately raises a question as to the enforceability of the condition. Accepting that a planning authority have extensive powers of entry, it none the less remains questionable whether a room containing a bed that is used diurnally for purposes other than sleeping – say as a study – might remain a "bedroom". Indeed, would it be sufficient that the room contain a bed, D even if the bed were not in regular nocturnal use for sleeping? That apart, it will generally be difficult to know whether the use of a room within domestic premises as a bedroom has ceased "permanently". A declaration by one proprietor respecting his intentions of using the room as a bedroom would not bind his successor in ownership of the property. Moreover, and perhaps more importantly, as counsel for the appellants pointed out, given E that the structure of which the planning authority complains affects the character of the building, by particular reason of its effect upon the floor plan and the classical proportions of what is a "principal" room, it is difficult to see how the particular domestic use, or mix of domestic uses, of the remainder of the principal room can be relevant to the question of whether consent should be given for the alterations. For these reasons, F among others, I consider this subhead of the condition to be flawed.

[21] While, on one view, the second subhead (net curtains) does not present perhaps the same enforcement problem, since the presence of such curtains is observable from the outside of the premises, it none the less also presents difficulties. There are the problems of drafting to which counsel for the appellants adverted. More fundamentally, there is the issue of G whether, in the context of the grant of a planning consent or a listed building consent, it is reasonable and lawful to require the deployment of soft furnishings, such as drapes or curtains, which do not form part of the immovable property. Planning conditions run with the land. Listed building consents similarly run with the land. If this condition were valid, it would follow that anyone purchasing the house from the present owner H must separately acquire the net curtains, or at least net curtains of the same (unspecified) type and dimensions as those present at the time of the reporter's visit. Even assuming in principle that an internal alteration to a principal room within a listed building that adversely affects the character of the building might receive consent if the windows were screened to

A prevent view from the outside, the terms of this subhead of the present condition do not admit alternatives, such as venetian blinds or any other screening device. On that ground, also, the present subhead of the condition is unsatisfactory. I have also to say that I have considerable reservations as to whether an alteration forming part of the heritable or immovable property may properly be given consent subject to conditions
B relating to the use of soft furnishings such as curtains, since planning conditions and listed building conditions relate essentially to immovable property. However, given the limited discussion before me, I refrain from expressing a concluded view.

[22] The third subhead relates to the removal of the privet hedge. While the hedge may be a less ephemeral feature of the state of the property and
C its contents than the net curtains viewed by the reporter, it is, I think, evident that, at the least, this condition, or subhead, suffers the problems of drafting and expression to which counsel for the appellant authority adverted.

[23] Even if that third subhead might, as a matter of validity, survive those criticisms, it is evident that, in judging the visibility of the structure
D to passers-by, the reporter treated the net curtains and the hedge in conjunction. He also treated use as a bedroom as a material factor. In these circumstances, I do not consider it possible to divide up, or sever, the various subheads of the reporter's condition, all of which appear to have been a material part of his decision.

[24] I am conscious that many of the criticisms of the reporter's
E decision in respect of the conditions to which he made it subject may seem, at first sight, to be matters of drafting rather than substance. It is, of course, important that those granting planning consent subject to conditions give care to the draftsmanship involved in framing such conditions. As the guidance to which I was referred points out, such conditions must be framed with care and precision. However, in the present case, what might
F appear as drafting matters are, in my view, a symptom of a more fundamental issue, which I put to counsel for the respondents in the course of the discussion before me. That underlying problem is that, while purporting to decide whether listed building consent for the works in issue should be granted, the reporter appears in fact to be addressing a different question, namely whether, the works having been erected
G without a prior grant of listed building consent, it was reasonable that they should now be removed. Hence, at a central point in his reasoning, one finds this sentence in para 21 of the decision letter:

> However, while the integrity of the appeal property as a listed building has been compromised, I am not convinced that this is so great as to justify the *immediate* removal of the offending structure.

H

The emphasis is mine, but the notion of not requiring "immediate" removal is then carried forward, first into the decision that "an unrestricted consent" is not appropriate, and thereafter into the apparent intention to effect a temporal restriction, other than the five-year "stay of execution" suggested by the owner of the property, by means of the

condition with its subheads, the criticisms of which I have discussed. A
What the reporter does not appear to have considered was whether, had an application been presented for consent for a future proposal to carry out the construction, within the "principal" room, of the bathroom and toilet and the surrounding partition walls, it would have been proper to grant consent, either outright or, in the context of the factors to which his decision refers, subject to conditions relating to the continuing particular B
domestic use of the remainder of the principal room, the continuous deployment of soft furnishings, and the maintenance of external planting. So viewed, I apprehend that, at least absent special circumstances in the appeal property, the reporter might have reached a different view.

[25] I appreciate that the reporter adopted what he recognised to be a "pragmatic" approach, and it is possible to sympathise with him to some C
extent in that approach. I also appreciate that, in some cases, the distinction between asking oneself the question whether it is reasonable that an offending structure be removed, and asking oneself the question whether a proposal *ab ante* to construct such a structure should receive consent, may be a fine one. But, in my view, the two questions are truly different questions, and to apply the former test may risk deploying, in D
favour of the property owner who has carried out the development or alteration without the consent required by law, a practical advantage not enjoyed by those owners more careful or scrupulous in their attention to the requirements of the legislation. It also invites the imposition of inappropriate conditions directed, not to the grant of consent, but to the cessation and removal of an unauthorised alteration to a listed building for E
which consent might never have been granted *ab ante*. To apply the former test is, in my view, an error in law.

[26] For the foregoing reasons, I must respectfully conclude that, in addressing ground (e) of the property owner's appeal to him, the reporter did misdirect himself in law in a manner that requires that his decision be quashed.

Appeal allowed.

R (on the application of Kides) *v* South Cambridgeshire District Council and others

[2002] EWCA Civ 1370

COURT OF APPEAL
Aldous, Laws and Jonathan Parker LJJ
9 October 2002

A **Material considerations – Section 70(2) of Town and Country Planning Act 1990 – Nature and extent of statutory duty to have regard to material considerations – Five-year period between resolution to grant planning permission and issue of planning decision – Material changes during five-year period – Whether council obliged to have regard to change of circumstances – Whether council in breach of statutory duty under**
B **section 70(2)**

In December 1995, the respondent council resolved to approve an outline planning application submitted by the interested parties for the comprehensive development of a 30ha site. In October 2000, the council issued a decision notice on the 1995 application, granting planning
C permission pursuant to the 1995 resolution. The appellant challenged that decision, contending, *inter alia*, that the council had failed to have regard to material considerations, pursuant to section 70(2) of the Town and Country Planning Act 1990. The appellant alleged that, between the date of the resolution and the grant of the planning permission, there had been material changes in the circumstances that the council were under a
D statutory duty to consider, including the introduction of PPG 3 in 2000. The appellant appealed the decision of the judge in the court below, dismissing her application for judicial review. The council and the interested parties contended that: (i) the council had discharged their duty under section 70(2); (ii) the appellant did not have *locus standi*; and (iii) the appellant had failed to apply for judicial review with promptitude.

E ***Held*** The appeal was dismissed.

An authority's duty to "have regard to" material considerations is not to be elevated into a formal requirement that, in every case where a new material consideration arises after the passing of a resolution (in principle) to grant planning permission, but before the issue of the decision notice, there must be a specific referral of the application back to the committee.
F The duty is discharged if, as at the date upon which the decision notice is issued, the authority have considered all material considerations affecting the application: see [122]. The test of a "material consideration" is an objective one. It is not for the delegated officer to decide what constitutes a material consideration within the meaning of section 70(2). Hence, it is no defence to a claim that an authority breached their section 70(2) duty
G for the authority to assert that, in issuing the decision notice, the delegated officer did not consider the consideration to be "material": see [127]. The

council were fully aware of the five new factors upon which the appellant A
relied, and they considered them with the 1995 application specifically in mind: see [129]. The appellant was entitled to present her challenge upon all available grounds: see [134]. She had lived in the area for upwards of 30 years, and had a real and genuine interest in seeking to prevent the substantial development permitted by the planning permission.

B

Cases referred to in the judgment

R (on the application of Burkett) v *Hammersmith and Fulham London Borough Council; sub nom R* v *Hammersmith and Fulham London Borough Council, ex parte Burkett; Burkett, Re* [2002] UKHL 23; [2002] 1 WLR 1593; [2002] 3 All ER 97; [2002] 2 PLR 90

R v *Monopolies and Mergers Commission, ex parte Argyll Group plc* [1986] 1 WLR 763;
[1986] 2 All ER 257, CA C

R v *Somerset County Council, ex parte Dixon* (1997) 75 P&CR 175; [1997] JPL 1030

R v *West Oxfordshire District Council, ex parte CH Pearce Homes Ltd* [1986] RVR 156; [1986] JPL 523

Appeal against a decision of Ouseley J

This was an appeal by the appellant, Erine Kides, against a decision of D
Ouseley J, refusing the appellant permission to apply for judicial review to quash a grant of planning permission, by South Cambridgeshire District Council, to Peter Sroude and Beazer Homes Central Ltd.

*Wyn Williams QC and Martin Edwards (instructed by Richard Buxton,
of Cambridge) appeared for the appellant, Erine Kides.* E

Alice Robinson (instructed by the solicitor to South Cambridgeshire District Council) represented the respondent council.

Richard Drabble QC (instructed by Marrons) represented the interested parties, Peter Stroude and Beazer Homes Central Ltd.

The following judgment was delivered.

F

JONATHAN PARKER LJ:

Introduction

[1] The issue on this appeal is whether South Cambridgeshire District Council (the council), as local planning authority, discharged their
statutory duty, under section 70(2) of the Town and Country Planning Act G
1990 (the 1990 Act), to have regard to all material considerations in dealing with an application for planning permission. That, in turn, raises an issue as to the nature and extent of the statutory duty.

[2] In the proceedings, Mrs Erine Kides (the appellant) applies for judicial review of a planning permission issued by the council, on
16 October 2000, for the comprehensive development of some 30ha of land H
at Home Farm, Longstanton, in Cambridgeshire. The planning permission was granted to Mr Peter Stroude and Beazer Homes Central Ltd (Beazer), respectively the major landowner and the developer of the site. They are joined in the proceedings as interested parties, and I will refer to them as such.

[3] Permission to apply for judicial review was refused by Collins J on the papers. At a renewed oral hearing, Sullivan J directed that the application for permission and the substantive application be heard together. The hearing took place before Ouseley J. By his order dated 26 October 2001, Ouseley J refused the appellant permission to apply for judicial review. However, in refusing permission, the judge delivered a reserved judgment, in which he set out the factual background and gave detailed consideration to the merits of the substantive application.

[4] By an order dated 20 June 2002, Sedley LJ granted permission to appeal against Ouseley J's order, and directed that the substantive judicial review hearing remain in the Court of Appeal, pursuant to CPR 52.15(4).

[5] The instant case is unusual, in that a period of almost five years had elapsed between the council's decision, in principle, to grant the planning permission and the issue of the decision notice granting permission. It is the appellant's case that a number of material changes in circumstances took place during that period, in the light of which it was incumbent upon the council, in discharging their statutory duty under section 70(2), to reconsider whether or not to grant planning permission; and that the council had failed to do that. She relies upon the following matters in particular:

1. Planning Policy Guidance Note 3 (PPG 3), published by the government in March 2000;

2. Government Circulars 13/96 (published in August 1996) and 6/98 (published in April 1998) relating to affordable housing;

3. The progress of the council's local plan review, with particular reference to housing allocations and density; and

4. The fact that, in May 1999, the airfield at nearby Oakington Barracks was declared redundant by the Ministry of Defence.

[6] The council and the interested parties contend that the council had discharged their duty under section 70(2), in that the members of the relevant committee were, at all material times, fully aware of the above matters, and that, in dealing with the 1995 application, the council had regard to them.

[7] Before the judge, the council and the interested parties contended, in addition, that permission should be refused on the ground that the claim form had not been filed within three months after the grounds to make the claim first arose, pursuant to CPR 54.5(1), submitting that the appropriate date for this purpose was a date prior to the issue of the planning permission. In the light of the subsequent decision of the House of Lords in *R (on the application of Burkett)* v *Hammersmith and Fulham London Borough Council* [2002] 1 WLR 1593[1] (reversing the decision of this court) that contention is no longer open to them, but they nevertheless maintain the contention (which they advanced in the alternative before the

[1] [2002] 2 PLR 90

judge) that permission to apply for judicial review should be refused as a A
matter of discretion on grounds of delay.

[8] The interested parties also challenged the appellant's standing to bring the proceedings.

Section 70 of the 1990 Act

[9] Section 70 of the 1990 Act is in the following terms (so far as B
material):

> 70.–(1) Where an application is made to a local planning authority for planning permission –
>
> (a) subject to sections 91 and 92, they may grant planning permission, either unconditionally or subject to such conditions as they think fit; or
>
> (b) they may refuse planning permission. C
>
> (2) In dealing with such an application the authority shall have regard to the provisions of the development plan, so far as material to the application, and to any other material considerations.

Planning history

[10] In order to set the appellant's challenge to the planning permission D
in its proper context, it is necessary to set out the planning history in some detail.

[11] The village of Longstanton lies some six miles to the north-west of Cambridge.

[12] In 1989, the Cambridgeshire structure plan identified Longstanton as a rural growth settlement. E

[13] In July 1991, following a public inquiry, the Secretary of State for the Environment dismissed Mr Stroude's appeal against the council's deemed refusal to grant planning permission for the comprehensive development of a 17ha site at Home Farm, Longstanton. The Secretary of State noted the inspector's finding that "the broad location of the development [was] a suitable and a natural extension to the settlement".
He also agreed with the inspector that a bypass for Longstanton was F
required on environmental grounds, noting also the view of the highway authority (Essex County Council) that "the acceptability and extent of the development of the [site] at Home Farm is totally dependent on the provision of the bypass". He agreed with the inspector that the proposals "could prejudice the resolution of a line for the bypass and that… they
represent an unacceptable extension of development into the open G
countryside". He accordingly concluded that planning permission should be refused.

[14] Between October 1990 and July 1991, a public inquiry took place at which representations were received concerning the draft South Cambridgeshire local plan. In April 1992, the inspector reported on those
representations. In para 5.50 of his report, the inspector said: H

> I regard the land at Home Farm as being in every way the best site for the bulk of the residential land allocated in Longstanton. It is therefore essential that its western limits should be defined so that it can be formally allocated in the Local Plan. This can only be done by preparation of a line for the

> A By-pass, which is something that the County council would not otherwise do in the near future. It is, however, agreed by them that the definition of the general route would not be unduly onerous, and I consider that this should be done as soon as possible so that the land with development potential could be defined. It would also have the advantage that the land to be safeguarded for the By-pass could be defined on the Inset Map and an appropriate safeguarding policy included in the Local Plan. This would, B however, only be acceptable if there is a good chance of the By-pass being constructed during the Plan period.

[15] The inspector accordingly recommended:

> that the County Council be asked to define a safeguarding line for the proposed Longstanton By-pass as soon as possible.

C

[16] In 1993, the local plan was adopted. It allocated two sites in Longstanton for residential development, one of which was at Home Farm. The site at Home Farm was intended to provide for up to 250 dwellings up until 2001. Other land was allocated for employment development. However, the local plan stated that substantial development D at Longstanton was dependent upon the prior completion of a bypass. It was contemplated that this might be secured under a section 106 agreement with the developer, but it was not at that stage contemplated that the developer would undertake to fund the entire cost of the bypass.

[17] In May 1995, the interested parties submitted an outline planning application (the 1995 application) for the comprehensive development of E the 30ha site at Home Farm, Longstanton, and for the construction of a complete bypass, the entire cost of which was to be borne by the developer.

[18] On 6 December 1995, the council's planning officer reported to the council on the 1995 application. The report recommended that the 1995 application be refused, on the ground (among other things) that a F commitment to the provision of some 250 dwellings beyond the year 2001:

> would prejudice and pre-empt the proper consideration and evaluation of all alternatives for future growth in the District during the statutory Local Plan review process.

[19] The report also noted, with some surprise, that Longstanton Parish G Council (of which the appellant was a member) endorsed the 1995 application unreservedly.

[20] At its meeting on 6 December 1995, the council rejected the planning officer's recommendation and resolved in principle:

> to approve [the 1995 application], subject to a report covering draft conditions and draft Section 106 Agreement being submitted for approval by H Committee at a future meeting.

[21] I will refer to this resolution as "the 1995 resolution".

[22] A report to the council's planning committee, dated 6 March 1996, noted that a first discussion draft of a section 106 agreement had been prepared and circulated, but stated that it was too early to advise members

of the progress of negotiations. The report recommended that the 1995 application be referred to the Secretary of State:

> along with all representations and a copy of this agenda report which summarises progress to achieve [the 1995 resolution].

[23] At their meeting on 6 March 1996, the planning committee resolved to refer the 1995 application to the Secretary of State as recommended, on the footing that it involved a departure from the local plan.

[24] On 29 March 1996, the Government Office for Eastern Region (the GOE) notified the council of the issue of an Article 14 direction to the effect that the council were not to determine the 1995 application while the Secretary of State considered whether to call it in. In a covering letter, the GOE said:

> The Article 14 direction has been issued, as it is considered that the significance of the development upon the village is such, that the determination of the planning application should be achieved in the context of the local plan review, which your Council is about to commence. The scale of the proposal indicates that it goes to the heart of the local plan's roll-forward of housing land provision to 2006 and its determination at this time could be considered to be premature and prejudicial to the local plan process.

[25] By a report dated 1 May 1996, officers reported the issue of the Article 14 direction to the planning committee.

[26] On 1 August 1996, the GOE notified the council that, after further consideration, the Secretary of State did not, after all, wish to intervene, and that the council were free to determine the 1995 application as they deemed fit.

[27] Later in August 1996, the first of the two government circulars relating to affordable housing (13/96) was published.

[28] On 4 September 1996, the Secretary of State's decision not to intervene in relation to the 1995 application was reported to the planning committee. The planning committee were also reminded that outline planning permission could not be issued pursuant to the 1995 resolution:

> until negotiations upon the Section 106 Agreement have been successfully concluded.

[29] In December 1997, the council published the consultation draft of the South Cambridgeshire Local Plan Review 1997. The review included "Policy Longstanton 1" in the following terms (so far as material):

> Two areas totalling 21ha are allocated for some 500 dwellings north of Over Road.
>
> Development of this site will be dependent upon the provision of a development related bypass secured through a local agreement. The agreement shall ensure that no more than 250 dwellings will be occupied before the bypass – including all necessary junctions and road links to the existing highway network is complete…

[30] On 8 April 1998, the second of the two government circulars relating to affordable housing (6/98) was published. The circular advised planning authorities and developers of the need to ensure that new development provided an appropriate mix of housing and other uses, and of different types of households. In its introduction, it advised that where there was evidence of a need for affordable housing, local plans should include a policy for seeking an element of such housing on suitable sites. Paragraph 10 of the circular set out a number of criteria to be applied by planning authorities in formulating policies for affordable housing to be included in local plans, and in assessing the suitability of sites to be identified in the plan and of any additional sites in respect of which allocations had not as yet been made.

[31] In May 1998, a development brief for the site was published and sent to a wide range of consultees. In December 1998, the development brief was adopted by the council as constituting supplementary planning guidance.

[32] In February 1999, the local plan was placed on deposit, in the form of the consultation draft, in order to give members of the public and other interested parties the opportunity to submit representations. In addition to Policy Longstanton 1, the draft local plan contained the following:

> 4.20 The [site at Home Farm] is not included within the requirements of affordable housing because it is the subject of a planning application where the Council has already resolved to grant planning permission under the provisions of the 1993 Local Plan without any requirements for affordable housing. At the time of the resolution, the Local Plan did not include a policy requiring affordable housing on allocated land and there was no affordable housing needs survey as required by government guidance. Any new planning application would not be subject to that resolution, and in the light of the high level of housing need identified by the District Wide Housing Needs Survey, it would be appropriate to consider whether affordable housing should be sought from this site. However, whether the site could stand the full 30% recommended by the survey, in addition to a bypass and other infrastructure, would need to be the subject of negotiations with the applicant.
>
> …
>
> 67.18 The District Council considers that the provision of the bypass is crucial for the village and therefore allocated a larger area for a housing estate than would otherwise be appropriate. In this instance there is no requirement for affordable housing as set out in Policy HG9 because of the need to ensure the bypass and other community facilities such as a village green, shop and surgery.
>
> …
>
> 67.20 The site of Home Farm, surrounded on three sides by existing housing, is considered appropriate for residential development. This would also enable the creation of a village green at the junction of Over Road and High Street. This, together with local shopping or community facilities which will be required because of the increased population, will allow the creation of a clearly identified centre to the village. However, the major area for housing development, as it will abut the bypass will require a high degree of

> screening and landscaping to reduce the visible impact and to create a satisfactory residential environment. …

[33] Thereafter, a large number of representations concerning the draft local plan were received from members of the public and other interested parties.

[34] Throughout this period, negotiations as to the terms of a section 106 agreement, in relation to the 1995 application, were still on foot. It appears that one of the causes of the delay in concluding a section 106 agreement was the fact that part of the site, the subject of the 1995 application, was owned by Cambridgeshire County Council, which were reluctant to enter into obligations involving a contribution to the cost of a bypass and associated road improvements, and which were looking to realise the maximum possible price for their land. For their part, the council were not prepared to allow planning permission to be issued, pursuant to the 1995 resolution, in the absence of a section 106 agreement in respect of all the land comprised in the Longstanton site. An attempt by Beazer to acquire the county council's land proved unsuccessful. The delays accordingly continued.

[35] At the same time, negotiations were also on foot between the council, the parish council, and the interested parties in relation to various ancillary matters, including the provision and maintenance of an area of public open space within the site.

[36] In March 1999, the interested parties, frustrated by the continuing delay in the issuing of planning permission pursuant to the 1995 resolution (a delay caused, as they saw it, by the attitude adopted by the county council), submitted a second planning application (the 1999 application) that excluded the council's land. The 1999 application sought permission for housing development over an area of 9.5ha, at an average density of not less than 25 dwellings per hectare. It also provided for the construction of stage 1 only of the proposed bypass, as a temporary measure (this being in contrast to the 1995 application, which, as noted earlier, provided for the complete bypass to be constructed at the cost of the developer).

[37] The council did not determine the 1999 application, and the interested parties accordingly appealed to the Secretary of State.

[38] At the beginning of October 1999, the local plan was published with an amendment relating to affordable housing that reflected the guidance contained in the two government circulars. Policy Longstanton 1 was also amended by the addition of a paragraph in the following terms (so far as material):

> Development of this site will be subject to Policies HG9… and HG10 for the provision of affordable housing. Where affordable housing is required, planning permission or renewal of planning permission will not be granted until a legal agreement has been signed ensuring such provision.

[39] On 6 October 1999, the planning director reported to the planning committee on the 1999 application. The report set out the background

A history, including the 1995 resolution, and referred to the council's policies in relation to affordable housing on allocated sites. After setting out the amended Policy Longstanton 1, the report went on to list a number of representations made by residents of Longstanton. Representation 2 in the report was in the following terms:

> B Empty houses will soon become available at Oakington Barracks. A greenfield site should not be developed in advance of that event or any other decision upon the future of the barracks. Brownfield development should take precedence. The large scale of potential development would suggest that it would provide for a significant amount of off-site infrastructure, which might include a Longstanton bypass.

C [40] Representation 10 was in the following terms:

> There have been material changes in circumstances since the resolution of the Council to approve the last application [a reference to the 1995 application]. These include:
>
> ● the emergence of the former airfield and barracks at Oakington/Longstanton as a likely area for large scale and comprehensive development within the
> D reasonably near future;
>
> ● the publication of new Government guidance, such as circulars 1/97 and 6/98, which require different benefits from large schemes, such as affordable housing provision;
>
> ● a third major change is the proposals themselves, which do not propose a bypass for Longstanton.

E [41] Representations 12, 13 and 14 were in the following terms:

> 12. The determination of the application would be premature in advance of the Public Inquiry into the Local Plan review, which would offer the appropriate forum to debate such a major issue.
>
> 13. The proposed development at Longstanton is a major departure from the adopted Local Plan and no justification has been given for this departure
> F in advance of the Local Plan Inquiry.
>
> 14. Any approval of the application would pre-empt the proper consideration which must be given to the Local Plan review on a district-wide basis.

[42] Under the heading "Planning Comments", the report said:

> G The application is now before the Secretary of State as an appeal for his decision. He has directed that the proposal does not require the submission of an Environmental Impact Assessment.
>
> The Council cannot issue a decision notice [ie in relation to the 1999 application]. But it should consider its position on the principal merits of the scheme in order to put its case at any future inquiry.
>
> The starting point is the Council's resolution of 6th December 1995 that it
> H was minded to approve a mixed development of 21 hectares housing (some 500 dwellings), 6.3 hectares business park, a B1050 bypass for Longstanton and related road works, extension to recreation ground (2.3 hectares), village green including land for local shop and surgery, open space, landscaping and related infrastructure. The Council considered that the provision of a bypass, financed by the development, outweighed the policy designations which

> would have limited the number of houses to be completed in the village to 250 allocated in the approved Local Plan 1993 and 400 houses normally permitted in a Rural Growth Settlement.
>
> The Secretary of State's decision of 1st August 1996 to allow the Council to determine the application as it deemed fit is of relevance.
>
> The outline application would have been approved by now had the interested parties and landowners reached a financial settlement which would have enabled the Section 106 Agreement to be signed.
>
> The Council needs to assess the new application against any material change in circumstances since those decisions on the earlier applications.
>
> The principal changes are:
>
> 1. the publication of the [Local Plan] and the representations made to that Plan;
>
> 2. the publication of Government guidance in Circular 6/98 encouraging authorities to include policies in local plans for seeking an element of affordable housing on suitable sites; and
>
> 3. the application itself, which does not propose a complete bypass for Longstanton and which proposes housing... on land not shown for such development on either the 1995 application or as part of a residential allocation in the Local Plan Deposit 1999.

[43] Then, after referring to the local plan, the report continued:

> Significantly, if the Regional Planning Guidance accepts the Panel's Report suggestion that a new settlement should be accommodated to the north, but close to Cambridge, the proposals will need to be examined through the Structure Plan process and ultimately into the Local Plan. So it would be premature to advance Oakington Barracks in isolation or in the context of a new settlement in this Local Plan Review...

[44] Under the heading "Affordable Housing", the report quoted para 4.20 of the local plan (see [32] above) and summarised the interested parties' representations on the issue of affordable housing as follows:

> 1. It is not necessary to provide additional affordable housing in Longstanton for the needs of Longstanton;
>
> 2. The community benefits from the provision of the Longstanton bypass and its capital costs to be financed from the development allocated for Longstanton outweigh the need to provide affordable housing...

[45] The report continued:

> In the consideration of the representations of the Deposit Local Plan it was suggested that any new planning application would not be subject to the earlier resolution (not to require affordable housing), and in the light of the high level of housing need identified in the District Wide Housing Needs survey, it would be appropriate to consider whether affordable housing should be sought from this site, though not necessarily the full 30% recommended by the survey.

[46] Turning to the merits of the 1999 application, the report commented that the fact that the application did not include the complete

A bypass represented a serious deficiency in the scheme, observing that the 1995 resolution was passed:

> because it would ensure the provision of the complete bypass at entirely the developer's expense...

[47] In conclusion, the report recommended that:

B (i) the council reject the 1999 application upon the grounds that: (a) it did not provide for a complete bypass; and (b) it proposed housing development in a location that had not been allocated in the local plan and that was poorly related to the remainder of the village and services within it; and

(ii) the council seek to secure a proportion of affordable housing based C upon housing needs, as part of a section 106 agreement.

[48] At the meeting of the planning committee held that day (6 October 1999), the recommendations in the report were adopted.

[49] On 8 October 1999, a special meeting of the council was held to assess the representations received concerning the draft local plan. In evidence is a tabulated document in which the various representations D are recorded, together with, against each representation, the council's assessment of the representation and any recommendation for a change in the plan in the light of that representation.

[50] One such representation was to the effect that the plan should include provision for affordable housing for local people on the Longstanton site. Complaint was made that the plan made provision for E affordable housing in relation to other, smaller sites, and that it was inequitable that the Longstanton site should be treated differently. The document records the council's assessment of this representation in the following terms (so far as material):

> The scale of development proposed is dependent upon the provision of a development related bypass. The District Council has resolved to F grant planning permission in respect of a 1995 application [a reference to the 1995 resolution] without the need for affordable housing. However, it would be appropriate to consider whether this provision should be sought from this site if any subsequent planning application were to be submitted... Such a provision would be subject to negotiation with the applicant...

G [51] In the light of this representation, the council resolved to change the plan by deleting the last sentence of para 67.18 (quoted earlier), and substituting a passage making it clear that any new planning application would not be subject to the 1995 resolution.

[52] Representations were also made as to the availability of land following the expected closure of the airfield at nearby Oakington H Barracks, which the Ministry of Defence had, in May 1999, declared to be redundant. The council's assessment of these representations, as recorded in the document that is in evidence, includes the following:

> The site at Oakington Barracks is of strategic importance, which the panel report identifies would require a long lead in time, clearly beyond the time

> scale of this local plan. It would be premature to identify this site [ie Oakington Barracks] in advance of the agreement of a Cambridge Sub Region Strategy beyond the period to 2006.

[53] The council recommended no change to the plan to reflect the representations relating to Oakington Barracks.

[54] In November 1999, the proposed changes to the local plan were published, and were the subject of further consultation.

[55] In March 2000, the government issued PPG 3[1]. For present purposes, it is necessary to refer only to paras 31, 32 and 38 of PPG 3, which were in the following terms (so far as material):

> 31. In deciding which sites to allocate for housing in local plans and UDPs [unitary development plans] local planning authorities should assess their potential and suitability for development against each of the following criteria:
>
> - **the availability of previously-developed sites** and empty or underused buildings and their suitability for housing use;
> - the **location and accessibility** of potential development sites to jobs, shops and services by modes other than the car, and the potential for improving such accessibility;
> - the **capacity of existing and potential infrastructure**...
> - the **ability to build communities**...
> - the **physical and environmental constraints on development of land**...
>
> **Allocating and Releasing Land for Development**
>
> 32. In determining the order in which sites identified in accordance with the criteria set out in paragraphs 30 and 31 should be developed, the presumption will be that previously-developed sites (or buildings for re-use or conversion) should be developed before greenfield sites. The exception to this principle will be where previously-developed sites perform so poorly in relation to the criteria listed in paragraph 31 as to preclude their use for housing (within the relevant plan period or phase) before a particular greenfield site...
>
> **Determining Planning Applications**
>
> ...
>
> 38. In considering planning applications for housing development in the interim, before development plans can be reviewed, local authorities should have regard to the policy contained in this PPG as material considerations which may supersede the policies in their plan... Where the planning application relates to development of a greenfield site allocated for housing in an adopted plan or UDP, it should be assessed, and a decision made on the application, in the light of the policies set out in this guidance. Comparison with available previously-developed sites against the criteria in paragraph 31 and in the light of the presumption in paragraph 32 and the policies on design, layout and efficient use of land, including car parking, will be particularly relevant. Where a proposed housing development involves the use of a previously-developed site or the conversion of existing buildings,

---

[1] Housing (2000)

> the proposal may need to be amended in accordance with this guidance, for example, in relation to design, layout, density and parking.

[56] At a meeting on 12 April 2000, the planning committee noted that, in the light of PPG 3, a review of housing allocations in the local plan would have to be undertaken as further preparation for the local plan inquiry.

[57] On 3 May 2000, the council's head of legal services reported to the planning committee on the progress that had been made in negotiating a section 106 agreement relating to the 1995 application. The particular issue that had arisen at that stage was whether the council were willing to assume the obligation of maintaining an area of public open space within the proposed development, should the parish council decide not to do so. In the introduction to the report, the head of legal services reviewed the history of the matter, as follows:

> Negotiations hitherto have proved very difficult indeed and although the Parish Council has instructed solicitors to assist it in its broad deliberations, it is not clear that negotiations in the normal form will be completed and a planning permission issued in the foreseeable future. The development is important for Longstanton, not least because it will afford the provision of a bypass which is much needed but also because it will advance the housing provision in our district. This matter has gone on since 1995 and has been held up by what have been unsuccessful negotiations with the County Council as part landowner. Those negotiations have now at last been concluded and the County Council has decided to sell its interest in the relevant land to developers (subject to planning permission) thereby allowing the whole project to go forward...

[58] The report recommended that the council should agree to adopt the public open space should the parish council be unwilling to do so. The recommendation was duly adopted by the planning committee.

[59] At their meeting on 7 June 2000, the planning committee received a verbal report from the head of legal services on the progress of negotiations with the parish council concerning maintenance of the area of public open space. The report was to the effect that negotiations were continuing but that a decision was expected shortly. The minutes record the members' concern at the prospect of any further delay. The relevant passage in the minutes reads as follows:

> Members welcomed the negotiations but expressed concern that extended discussions might significantly delay the process. The Head of Legal Services reiterated that a decision was expected quickly and that the issue would be returned to Committee in July if negotiations proved unsuccessful.

[60] At the July meeting of the planning committee (which took place on 12 July 2000), the committee had before them a document entitled "A Study of [PPG 3] on The Local Plan Review within the Context of the Approved Cambridgeshire Structure Plan Review 1995". I will refer to this document as "the study". Also before the committee was a report on the study by the planning director. It is to be noted that although the

committee are referred to, in this context, as the "planning policy A
committee", the evidence is that the planning policy committee were in fact the same committee as the planning committee, albeit under another name.

[61] The study contained a district-wide review and appraisal of housing allocations. Paragraph 23 of the study, under the heading "Site Assessment Criteria (Physical and Environmental Constraints)" was in the B
following terms (so far as material):

> Paragraph 32 of [PPG 3] makes it clear that whilst brownfield sites should be developed before greenfield sites, this should not be so where brownfield sites perform so poorly in relation to the criteria in paragraph 31. In the context of this study it is considered that sites in unsustainable villages [which included Longstanton: see below] fall into this category of poor C
> performance...

[62] Paragraph 30 of the study, under the heading "Conclusions from the Analysis – Selection of Sustainable Locations", was in the following terms (so far as material):

> ... There are a small number of villages where the Local Plan makes D
> allocations in villages which do not meet the "sustainable" criteria. In each case there are specific reasons justifying the proposed development and these accord with criterion (4) of paragraph 31 of [PPG 3]... Thus:
>
> ...
>
> ● Longstanton, a Rural Growth Settlement designated in order to provide a much needed bypass, together with an extension to the recreation ground, E
> a village green and a shop...

[63] Paragraphs 37 to 40 of the study dealt with the question of density. In para 39, it was proposed that, for those sites where there were no identified constraints, the notional average capacity for housing allocations be raised from 25 dwellings per hectare to 30 dwellings per hectare. Paragraph 40 of the study was in the following terms (so far as F
material):

> This should not preclude the use of a lower (or higher) density if local circumstances dictated that there are material considerations to do so. These are:
>
> ...
>
> ● Longstanton where the local plan specifies a maximum of 500 dwellings G
> within the agreed line of the proposed bypass and where planning permission has been agreed in principle [a reference to the 1995 resolution] and negotiations on the required Section 106 Agreement are close to completion.

[64] Under the heading "Conclusions", paras 57 and 58 of the study H
were in the following terms:

> 57. PPG3 requires local planning authorities to consider increasing the density of new housing developments. Whilst densities of the order of 30-50 dwellings per hectare may be appropriate in urban situations, an average density of 30 dwellings per hectare is appropriate in most rural situations.

A The conclusion is therefore drawn that average notional density in South Cambridgeshire should be increased from 25 to 30 dwellings per hectare.

> 58. The suggested approach of increasing housing density and a revised assessment of windfalls leads to the conclusion that there should be a reduction in the overall level of housing allocations, and a number of sites are identified which could be deleted to bring the overall level to that which is necessary to meet the Structure Plan requirement.

B

[65] In his report on the study, the planning director informed the planning committee that the inspector conducting the local plan inquiry had arranged a number of "round table" sessions, at which objectors could discuss relevant issues informally with the inspector and representatives of the council. He further informed the committee that the inspector had C identified a number of specific issues to be debated, including:

> Should existing allocations be reviewed?

[66] The report went on to recommend that the inspector be advised (among other things) that the view of the council as to density was that, D unless local circumstances should dictate otherwise, the notional average housing density for sites allocated in the local plan be increased from 25 to 30 dwellings per hectare in accordance with PPG 3, and that, in view of the increased density permitted by PPG 3, a number of sites that were allocated for housing development in the local plan be deleted. Seven sites were then listed. The list did not include the site at Home Farm, which was the subject of the 1995 application.

E [67] These recommendations were duly adopted by the planning committee at its meeting on 12 July 2000.

[68] The recommendations of the planning committee came before the full council at their meeting on the following day, 13 July 2000. The council adopted the recommendations of the planning committee, save that they F reduced the list of allocated sites to be deleted from the local plan from seven to four.

[69] At a meeting of the planning committee on 2 August 2000, the planning director reported, in relation to the 1995 application, that it was likely that agreement would shortly be reached with the parish council concerning the public open space, and that negotiations in relation to the G long-outstanding section 106 agreement were at an advanced stage. It was accordingly anticipated that planning permission would be issued within the next few days, pursuant to the 1995 resolution. The minutes record that "members approved this approach".

[70] On 16 October 2000, the decision notice on the 1995 application was issued, granting outline planning permission pursuant to the H 1995 resolution.

[71] Finally, so far as the planning history is concerned, I should record that throughout the period from the passing of the 1995 resolution, on 6 December 1995, until the issue of the planning permission, on 16 October 2000, it was the practice of the planning director to place before the planning committee, at quarterly intervals, a list of all outstanding

planning applications. These lists would have included the 1995 application. A

Arguments before the judge

[72] In summarising the argument before the judge, I omit reference to the issue of whether the claim form was filed within three months after the grounds to make the claim first arose (see CPR 54.5(1)), since, as explained earlier, that issue was put to rest by the decision of the House of Lords in *Burkett*. B

Appellant's contentions

[73] As a matter of law, it was contended on behalf of the appellant, relying upon R v *West Oxfordshire District Council, ex parte CH Pearce Homes Ltd* [1986] JPL 523, at p527 *per* Lord Woolf, that, in discharging their duty under section 70(2), a local planning authority are required to reconsider a resolution to grant planning permission if there is a change in material considerations occurring after the passing of the relevant resolution. C

[74] As to the facts, it was contended on behalf of the appellant that PPG 3 marked a radical change in government policy with regard to the location and design of housing sites, with an emphasis away from greenfield development. Reliance was also placed in this regard upon the two government circulars relating to affordable housing. In the light of these changes in government policy, a housing scheme considered in 2000 would, it was submitted, be subject to very much more stringent requirements, in relation to density and provision for affordable housing, than those that obtained in 1995. It was also contended that it was a material consideration that, during the period following the 1995 resolution, the Oakington Barracks site, a brownfield site, had become available for early housing development. An additional material consideration, it was submitted, was whether a grant of planning permission, pursuant to the 1995 resolution, would be premature, in that it would prejudice the outcome of the local plan review. D E F

[75] These changes, it was submitted, were sufficient to trigger the council's duty to reconsider the 1995 resolution, yet the planning committee were never asked to, nor did they, undertake such a reconsideration.

[76] On the question of promptness, it was submitted that, although the proceedings were not issued until some six weeks after the issue of the planning permission, the appellant lost no time in instructing a specialist solicitor, Mr Richard Buxton, who in turn communicated promptly with the council in order to establish whether there was a proper basis for judicial review proceedings. Reliance was also placed in this connection upon the enormous volume of documentation involved. In the circumstances, it was submitted, the appellant could not have been expected to have acted more expeditiously than she did. G H

Interested parties' contentions

[77] The interested parties submitted, as a preliminary point, that the appellant had no standing to bring the proceedings, since she relied,

among other things, upon the absence of any provision in the 1995 application for affordable housing, which was not a matter of any real concern to her.

[78] As to the merits, the interested parties accepted, as they were bound to do, that, in dealing with the 1995 application, the council were under a duty to have regard to all material considerations, but it was submitted that this did not mean that an officer, to whom had been delegated the authority to issue the consent, was obliged to refer the matter back to committee whenever a change of circumstances, however small, occurred, and regardless of the views expressed by the planning authority in other contexts. It was submitted that the officer to whom such a power has been delegated is entitled to act upon the resolution, provided that there has not been a change of circumstances of a kind that means that the relevant committee should be invited to form a fresh judgment. In the instant case, it was submitted, it was plain that the planning officer's mandate remained good, and that no question of referral arose.

[79] As to affordable housing, the interested parties pointed out that the council specifically addressed this consideration in the context of the 1999 application. As to density, this too was a matter specifically considered by the council in 2000, in the context of the local plan review. In the result, the council had decided not to alter the density of the Home Farm site or to delete it from the local plan. On the question of prematurity, it was submitted that the council were fully aware of the scale of the proposed development and its relationship with the emerging local plan, through the involvement of members of the planning committee with matters of policy (the planning policy committee being the planning committee under another name). As to Oakington Barracks, it was submitted that any development of the airfield was some way off. Reference was made, in this connection, to the fact that the Ministry of Defence had made it clear that the development of the site required "positive outputs from a number of relevant studies to be progressed".

[80] As to promptness, it was submitted on behalf of the interested parties that there had been substantial delay and inaction on the part of the appellant, for which she was entirely to blame, and that, in consequence, the interested parties had suffered prejudice. It was accordingly submitted that relief should, in any event, be refused as a matter of discretion.

Council's contentions

[81] The council's contentions were substantially the same as those advanced on behalf of the interested parties, save that the council did not take any point as to the appellant's standing to bring judicial review proceedings.

Judgment of Ouseley J

[82] After setting out the planning history, the judge turned, in [65] of his judgment, to the issue as to the nature of the council's duty in determining the 1995 application. He concluded that the duty continued up to the moment when the decision notice was issued, and that the

discharge of that duty might, depending upon the facts of the particular A case, require that an earlier decision be reconsidered. However, in [66] of his judgment, he rejected the submission made on behalf of the appellant that only a formal reconsideration can ever suffice for compliance with section 70(2), in the event of new material considerations arising between a resolution to grant planning permission and its actual grant.

[83] In [69] of his judgment, the judge said: B

> In my judgment it is a question of fact whether in the circumstances in any particular case the duty has been fulfilled. What actual steps have to be taken in order to fulfil it, as a matter of fact, depends on the circumstances of the case. Very often the circumstances will indeed require a formal Committee meeting to reconsider an earlier resolution in the light of new circumstances. However I do not accept that that is required in law if, on the facts of a C particular case, it is possible to demonstrate that regard was had to material considerations in some other way.

[84] In [71] of his judgment, the judge rejected the contention, advanced primarily by the interested parties, that the relevant consideration was whether there was any good reason for the officer with delegated D authority to issue the planning permission to refer the matter back to the planning committee. In relation to that contention, the judge said:

> The delegation of the consideration of new material considerations is not an answer to the Appellant's claim: first, no resolution generally or particularly delegating the necessary powers of determination was produced to me; and, secondly, a power to determine whether an application should be brought E back to Committee and to deal with mechanical or trivial issues which might reasonably not require Committee consideration, is very different from a power to determine an application in these circumstances, where new factors of obvious significance had arisen.

[85] In [72] of his judgment, the judge observed that there was no F basis for supposing that members of the council were unaware that the 1995 resolution was not binding on the council, or that it disentitled the council from revisiting the question of whether planning permission should be granted.

[86] As to whether, on the facts, the council had discharged their statutory duty in relation to the 1995 application, the judge G concluded that there could be no doubt that the members of the planning committee were well aware of the contents and significance of PPG 3, with its emphasis upon priority being given to brownfield sites and on higher density for housing development. Moreover, he commented that at no time did members express a desire to change their minds about granting the 1995 application, as they could easily have H done.

[87] As to the debate concerning the local plan, the judge, in [74] accepted that this took place in a different context from that of the 1995 application, but he went on to observe that the objections to the local plan, that were considered by the council, were:

> inevitably objections to the development itself; consideration of them as policy objections necessarily involved consideration of the principle of the development…

[88] In [77] of his judgment, the judge concluded that the 1995 application was "inextricably intertwined" with policy issues relating to the local plan and to the allocation of sites for housing development. He further concluded that the objections were:

> considered and weighed on their merits and not dealt with as if, however sound they might be, the [1995] resolution had sold the pass…

[89] The judge then considered in turn the specific changes relied upon by the appellant.

[90] As to PPG 3, the judge concluded that the council's consideration of PPG 3 went beyond a general awareness of its existence, and that, in their analysis of the planning director's report on PPG 3, the council clearly took account of the competing considerations relating to development at Longstanton.

[91] As to density, the judge said, in [80]:

> Density was specifically considered in relation to Longstanton and an exception to the general approach was indicated, not because of the 1995 resolution, but because of the purpose of the housing development and its relationship to services and the bypass.

[92] The judge went on to point out that density was, in any event, a reserved matter, which was not covered by the grant of outline planning permission. He continued, in [80]:

> The consideration of [PPG 3] in the context of this study inevitably involved a consideration of the 1995 application – the maintenance of the allocation was not the consequence of the 1995 resolution but was the consequence of the consideration by the Council and its desire to see the development go ahead with the advantages which it would bring, some of which were relevant to PPG 3 particularly in the context of a rural district with few brownfield sites available.

[93] As to Oakington Barracks, the judge concluded, in [81], that:

> Oakington Barracks was not rejected as an objection because the [1995 resolution] precluded consideration being given to it; it was rejected because it did not have sufficient merit to warrant the allocation being deleted and hence to warrant planning permission being refused.

[94] As to affordable housing, the judge concluded, in [82], that the policies set out in PPG 3, and in the two government circulars, were considered by the council in relation to the 1995 application, and that a view was reached that "the current package would not be disturbed". He continued, in [83]:

> the substance of the consideration involved both the current and any potential fresh application and the substance of the conclusion was that the

current application now enjoyed an agreed package in which the question whether or not to have affordable housing had been considered and rejected because of other and high development costs. A

[95] As to prematurity, the judge observed that the planning committee were obviously very well aware, throughout 2000, that the local plan inquiry was under way, that there were objections to the local plan, and that the allocation at Longstanton was very important to the plan and to Longstanton itself. He continued, in [85]: B

> It is clear that the Planning Committee wanted the development rapidly to proceed. It obviously rejected the very notion that the grant of permission in 2000 following a resolution in 1995 was premature in relation to a Local Plan which was more advanced than it had been in 1995. C

[96] The judge went on to note that, in considering the 1999 application, the planning committee were specifically alerted to the question of prematurity, to the possible future availability of Oakington Barracks, and to the question of affordable housing.

[97] In [87] and [88] of his judgment, the judge said: D

> 87. I accept that there was not one single meeting at which the new material considerations were formally considered as a group together with other relevant factors explicitly in relation to the 1995 application. This might matter if the Planning Committee had accepted the strength of individual objections on separate occasions which individually did not justify a change in decision, but which cumulatively might have done had they been so considered. However, here the Council did not accept that any of the new factors were justified or significant objections at all, let alone ones which taken together could lead to a different result. E
>
> 88. I consider that the Council did fulfil its section 70 duty when granting planning permission in relation to all those material considerations. It considered them principally but not exclusively through the local plan process but it clearly maintained its view having considered each of them. F

[98] The judge accordingly rejected the application on its merits.

[99] The judge then turned, in [90] to [100] of his judgment, to the question of delay. In this part of his judgment, the judge addressed the issue as to when time started to run, pursuant to CPR 54.5(1). As explained earlier, that is no longer a live issue in the light of the House of Lords decision in *Burkett*. In the result, the judge concluded that the proceedings were issued out of time. Accordingly, it was unnecessary for him to address the alternative argument based upon delay, viz that relief should be refused as a matter of discretion. G

[100] The judge then turned, in [101] to [110] of his judgment, to the question of the appellant's standing to bring the proceedings. He rejected the interested parties' submission that the appellant had no standing at all, but he accepted a narrower submission that, in relation to the issue as to affordable housing, she was "a mere meddler" and had no standing. He concluded, in [109], that the appellant had no interest at all in securing H

A provision for affordable housing, and that she had seized adventitiously on a point of no interest to her. He accordingly concluded that it would be an abuse of process for the argument based upon affordable housing to proceed, and that permission to proceed with that particular argument should be refused.

[101] Finally, under the heading "Discretion", the judge said, in [111]:

B

> If I had granted permission, and if the affordable housing ground had been the basis of success for the appellant as a matter of law, I would have refused her relief in the exercise of my discretion for the reasons which I have given in relation to standing.

Arguments on this appeal

C [102] For the appellant, Mr Wyn Williams QC submits that, on the true construction of section 70(2), the council were required to give focused consideration of the 1995 application in the light of all material considerations, as opposed to what he described as tangential consideration, where material considerations were considered in different contexts. He does not go so far as to submit that there must in every
D case be a separate meeting, at which the application in question is considered. Rather, he submits that there has to be a process in which the court and the parties affected can be confident that all material considerations have been considered, both individually and collectively, and that a balancing process had been undertaken. Anything short of that will, he submits, fall short of what the statute requires. In
E practical terms, he submits, the sensible course is to hold a meeting at which all considerations material to a particular application can be considered in the specific context of that application. He submits that the statutory duty continues up to the point when the decision notice is issued. He points out the only meeting at which the 1995 application was considered in the light of all material considerations was the meeting on
F 6 December 1995, at which the 1995 resolution was passed, and that no further such meeting took place prior to the issue of planning permission on 16 October 2000.

[103] He submits that there are good policy reasons for adopting such an approach to the construction of section 70(2), in that the construction for which he contends would be conducive to good
G and consistent decision making and would serve to promote transparency in the decision-making process. He submits that it is highly unsatisfactory that it was necessary for the judge in the instant case to, in effect, carry out a paperchase in order to determine whether the council had discharged their statutory duty. Members of the public ought to be able to see for themselves what the decision-making process
H was.

[104] Turning to the wording of section 70(2), Mr Williams submits that the expression "[i]n dealing with..." must relate back to the planning authority's functions under subsection (1). Thus, since the act of the council in considering objections to the local plan is not an activity within subsection (1), it cannot constitute "dealing with" a planning application.

[105] He submits that section 70(2) requires that there should be a thought process that is focused specifically upon the application that is being dealt with. A

[106] Turning to the facts of the instant case, Mr Williams submits that at no stage did the council engage in the requisite thought process. When new material considerations emerged (for example, in PPG 3, or in the circulars relating to affordable housing), they were considered piecemeal B and sequentially, and not specifically in the context of the 1995 application. On the question of prematurity, Mr Williams submits that that question, too, fell to be considered as part of the ongoing planning process and the continuing evolution of the local plan. He submits that the various threads were never drawn together, with the consequence that the point was never reached, in the decision-making process, where the decision maker was in C a position to have regard to all material considerations, to weigh them up and to take a decision based upon them.

[107] Ms Alice Robinson, for the council, submits that the expression "[i]n dealing with", in section 70(2), should not be restricted to consideration of the application in question by the relevant committee. Rather, she submits, it embraces the totality of what the planning authority D say and do, in so far as it is relevant to the application in question, in the sense that it has a direct bearing upon it.

[108] On the facts, Ms Robinson supports the judge's conclusion that the council had regard to all material considerations in relation to the 1995 application, despite the fact that there was no specific reconsideration of the 1995 resolution. She points out that, for over E 10 years prior to the grant of planning permission, the Home Farm site had been considered suitable for the provision of a substantial quantity of housing, and that the fact that the 1995 application provided for the construction of a complete bypass at the developer's cost was plainly considered by the council to be a considerable attraction. She also points out that the merits of the proposed development at Longstanton F continued to be debated by the council in the context of the local plan review. She further points out that the council were informed, on a quarterly basis, that the 1995 application remained outstanding, and that, in 2000, the 1995 application was twice considered specifically by the council in the context of the negotiations concerning the maintenance of the public open space. G

[109] As to affordable housing, Ms Robinson reminds us that, in the course of the local plan process, the council considered and assessed objections that the plan made inadequate provision for affordable housing, and concluded that, in relation to any new planning application, it would be appropriate to consider whether provision for affordable housing should be made, and that the plan be amended accordingly. H

[110] As to housing allocation policy, Ms Robinson relies upon the study (to which I referred earlier), that specifically considered the Longstanton allocation (which, in turn, reflected the 1995 resolution), and concluded that it should not be changed. The study was considered by the council in the context of formulating the council's position in the

forthcoming local plan inquiry, and the council accepted its recommendations.

[111] As to density, Ms Robinson points out that (as with the policy on housing allocations) the guidance on density provided by PPG 3 was considered in the study, and that, once again, the council accepted their officer's recommendations based upon the study, to the effect that, unless local circumstances dictate otherwise, the notional average housing density for allocated sites be increased from 25 to 30 dwellings per hectare. Local circumstances, for this purpose, included the fact that, at Longstanton, a maximum number of dwellings had been specified within the agreed line of the proposed bypass, that outline planning permission had been agreed in principle, and that the lengthy negotiations on the terms of a section 106 agreement were close to finality. Ms Robinson also points out that, in any event, matters of density were reserved by the 1995 resolution.

[112] As to Oakington Barracks, Ms Robinson points out that objections based upon the prospective availability of the Oakington Barracks site were considered by the council in the context of the Longstanton allocation, but that the council considered that the allocation should remain unchanged.

[113] As to delay, Ms Robinson submits that relief should, in any event, be refused on grounds of discretion, on the ground that the appellant delayed unduly before bringing the proceedings. She submits that it should have been clear to the appellant well before 16 October 2000 that the council were proposing to issue a decision notice, granting planning permission pursuant to the 1995 resolution, without a specific reference back to the council to reconsider the 1995 application in the light of the five matters upon which the appellant relies. She submits that had the appellant sought legal advice, and drawn the council's attention, at the proper time, to the material considerations to which the council allegedly failed to have regard, the council would have had the opportunity to correct the alleged defect by specifically reconsidering the 1995 application in the light of those considerations. In the event, her failure to do so has, she submits, caused the council unnecessary expense and delay. She further submits that if the planning permission is quashed, there will be further delay in the provision of much needed housing and of the Longstanton bypass.

[114] For the interested parties, Mr Richard Drabble QC essentially adopts Ms Robinson's submissions on the facts. In summary, he submits as follows:

1. In relation to PPG 3, he submits that the fact that the Home Farm site was not included in the list of sites where changes to the local plan were proposed is plainly indicative of the fact that the council had considered the Home Farm site in the context of PPG 3, and were of the view that there should be no changes in relation to that site.

2. As to Oakington Barracks, he submits that the position is substantially the same. The site at Oakington Barracks was not yet available for housing, and the prospect of its becoming available at some

time in the future was not considered to be a factor of sufficient weight to justify any change in the council's thinking, so far as the Home Farm site was concerned. A

3. As to density, he reminds us that this was, in any event, a reserved matter.

4. As to affordable housing, he submits that the evidence establishes that the council had regard to the two government circulars when considering the local plan, and that, in so doing, the council addressed the question of whether to require a provision for affordable housing as a condition of allowing the 1995 application but decided not to do so. B

5. As to prematurity, he submits that had the Oakington Barracks site been available in the near future as a viable alternative site a prematurity objection might have some force, but, on the evidence, the Oakington Barracks site was not expected to become available for some time. Hence it was, as he put it, kicked into touch. He further points out that prematurity is relied upon by the appellant as a reason for refusing planning permission; it is not contended that the council should have merely adjourned further consideration of the 1995 application to enable them to wait and see how matters turned out. C D

[115] As to section 70(2), Mr Drabble submits (as he submitted unsuccessfully to the judge: see [71] of the judgment, quoted earlier) that, where there has been a material change of circumstances between the resolution and the issue of the decision notice consequent upon the resolution, the relevant question is whether the officer with authority to issue the decision notice remains entitled to act upon the resolution, and that he will remain entitled so to act provided that there has not been a change of circumstances "of a kind that means the committee should be invited to form a fresh judgment". Paragraph 12 of the skeleton argument continues: E

> He is not obliged to go back to committee if his mandate remains good, as it will if the changes are small or if the original resolution remains consistent with the policy views of the council, albeit expressed in a different context. If it is clear that the council as a whole (including the officers holding delegated powers) are alive to the various changes of circumstances, the council will have regard to all material considerations at the date of issue; express reconsideration by the original committee is not necessary. F

G

[116] As to discretion, Mr Drabble adopts the submissions of Ms Robinson as to delay. He submits that the delay has caused prejudice to the interested parties, in that: Mr Stroude has refrained from investment in Home Farm and its buildings in the expectation that permission would, in due course, be granted pursuant to the 1995 resolution; Beazer has required a full archaeological survey to be carried out on the site, which interfered with drainage and soil quality; stocks of grain have had to be disposed of, thereby seriously constraining his agricultural operations; and Beazer has incurred substantial expense, amounting to approximately £500,000, in commissioning studies and appointing a number of consultants and advisers. H

[117] Mr Drabble also makes the further submission (accepted by the judge, in the context of affordable housing: see [109] of the judgment, quoted earlier) that the appellant lacks standing in relation to the proceedings, since she has taken objections to the planning permission on grounds that are, in themselves, of no real concern to her. Mr Drabble submits that the appellant is a "mere meddler": see *R* v *Somerset County Council, ex parte Dixon* [1997] JPL 1030, at p1037.

Conclusions

[118] I begin by considering the nature and extent of a planning authority's duty under section 70(2) of the 1990 Act.

[119] Section 70(2) requires a planning authority, in "dealing with" an application, to "have regard" (among other things) to all "material considerations".

"Dealing with"

[120] In the context of the activities of a planning authority in relation to a planning application, I find it hard to think of an expression that has a wider or more general meaning than the expression "[i]n dealing with". In my judgment, "dealing with", in the context of section 70(2), includes anything done by, or on behalf of, the planning authority that bears in any way, and whether directly or indirectly, upon the application in question. Thus, it extends beyond "considering", so as to include administrative acts done by the authority's delegated officers. Nor, in my judgment, is the expression "dealing with" to be limited to the particular acts of the authority in granting or refusing permission under section 70(1). I would regard such a construction as an unjustifiable limitation upon the natural meaning of the words. In temporal terms, the first act of a planning authority in "dealing with" an application will be their receipt of the application, and their final act will normally be the issue of the decision notice (certainly that is the position in the instant case).

"Material considerations"

[121] In my judgment, a consideration is "material", in this context, if it is relevant to the question as to whether the application should be granted or refused; that is to say, if it is a factor that, when placed in the decision-maker's scales, would tip the balance to some extent, one way or the other. In other words, it must be a factor that has some weight in the decision-making process, although, plainly, it may not be determinative. The test must, of course, be an objective one, in the sense that the choice of material considerations must be a rational one, and the considerations chosen must be rationally related to land-use issues.

"Have regard to"

[122] In my judgment, an authority's duty to "have regard to" material considerations is not to be elevated into a formal requirement that, in every case where a new material consideration arises after the passing of a resolution (in principle) to grant planning permission but before the issue of the decision notice, there has to be a specific referral of the

application back to committee. In my judgment, the duty is discharged if, as at the date at which the decision notice is issued, the authority have considered all material considerations affecting the application, and have done so with the application in mind – albeit that the application was not specifically placed before them for reconsideration.

[123] The matter cannot be left there, however, since it is necessary to consider what is the position where a material consideration arises for the first time immediately before the delegated officer signs the decision notice.

[124] At one extreme, it cannot be a sensible interpretation of section 70(2) to conclude that an authority is in breach of duty in failing to have regard to a material consideration, the existence of which they (or their officers) did not discover or anticipate, *and could not reasonably have discovered or anticipated*, prior to the issue of the decision notice. So there has to be some practical flexibility in excluding from the duty material considerations to which the authority did not *and could* not have regard prior to the issue of the decision notice.

[125] On the other hand, where the delegated officer who is about to sign the decision notice becomes aware (or ought reasonably to have become aware) of a new material consideration, section 70(2) requires that the authority have regard to that consideration before finally determining the application. In such a situation, therefore, the authority of the delegated officer must be such as to require him to refer the matter back to committee for reconsideration in the light of the new consideration. If he fails to do so, the authority will be in breach of their statutory duty.

[126] In practical terms, therefore, where, since the passing of the resolution, some new factor has arisen of which the delegated officer is aware, and which might rationally be regarded as a "material consideration" for the purposes of section 70(2), it must be a counsel of prudence for the delegated officer to err on the side of caution, and to refer the application back to the authority for specific reconsideration in the light of that new factor. In such circumstances the delegated officer can safely proceed to issue the decision notice only if he is satisfied that: (a) the authority are aware of the new factor; (b) they have considered it with the application in mind; and (c) on a reconsideration, the authority *would* reach (not *might* reach) the same decision.

[127] In substance, therefore, I accept the submission made by Mr Drabble in para 12 of his skeleton argument (quoted in [115] above), but with the proviso (that may in any event be implicit in his formulation of the statutory duty) that the test of a "material consideration" is an objective one in the sense explained in [121] above. It is not for the delegated officer to decide what is a material consideration within the meaning of section 70(2). Hence, it is no defence to a claim that an authority have breached their section 70(2) duty for the authority to assert that, in issuing the decision notice, the delegated officer did not consider the consideration to be "material". Accordingly, I respectfully agree with the judge's observation, in [71] of the judgment, that "[t]he delegation of

A the consideration of new material considerations is no answer to the… claim".

[128] Having identified the nature and extent of the council's duty under section 70(2), I now turn to the facts of the instant case in order to determine whether, on those facts, the duty was discharged.

[129] In my judgment, the lengthy recital of the planning history that
B I set out earlier in this judgment admits of only one answer to that question. It seems to me plain, on the facts, that not only were the council fully aware of the five new factors upon which the appellant relies, but they considered them (had regard to them) with the 1995 application specifically in mind, in that the 1995 application was one (and a prominent one) of a number of matters that together set the context in which the new
C factors were considered and assessed. In the light of the decisions that the council took, and the policies that they adopted, it is entirely clear, in my judgment, that had the planning officer taken it upon himself to refer the 1995 application back to committee for reconsideration immediately before issuing the planning permission, the council's decision would have been the same. Indeed, it goes further than that, in my judgment. Given
D the very considerable period that had elapsed since the passing of the 1995 resolution, and the understandable concern of members at the continuing delay in negotiating the terms of a section 106 agreement, the inference that I draw is that it would have come as a considerable and unwelcome surprise to members had they been told, at the beginning of October 2000, that although the section 106 agreement was finally in place,
E it was nevertheless necessary for them specifically to reconsider the 1995 application before a decision notice could be issued pursuant to the 1995 resolution.

[130] Accordingly, I consider that the judge was right to conclude (in [88] of his judgment) that, in granting planning permission pursuant to the 1995 resolution, the council discharged their duty under section 70(2).

F [131] In the light of that conclusion, the issue of discretion does not strictly arise. As to that, I would only say that had I concluded that relief ought otherwise to be granted to the appellant, I would have concluded that there were no good grounds for refusing relief as a matter of discretion. In particular, the suggestion that the appellant should have warned the council, before they issued the decision notice, that they were
G acting unlawfully seems to me to be fanciful.

[132] That leaves the issue of standing. As to that, it seems to me that there is an important distinction to be drawn between, on the one hand, a person who brings proceedings having no real or genuine interest in obtaining the relief sought, and, on the other hand, a person who, while legitimately, and perhaps passionately, interested in obtaining the relief
H sought, relies as grounds for seeking that relief upon matters in which he has no personal interest.

[133] I cannot see how it can be just to debar a litigant who has a real and genuine interest in obtaining the relief that he seeks from relying, in support of his claim for that relief, upon grounds (which may be good grounds) in which he has no personal interest.

[134] It seems to me that a litigant who has a real and genuine interest in challenging an administrative decision must be entitled to present his challenge on all available grounds. Nor do I read the judgment of Sedley J (as he then was) in *ex parte Dixon* as casting doubt upon that proposition. Similarly, Lord Donaldson MR's reference (in *R* v *Monopolies and Mergers Commission, ex parte Argyll Group plc* [1986] 1 WLR 763, at p773) to "the applicant's interest" is, as I read it, a reference to the applicant's interest in obtaining the relief sought: in this case, the quashing of the planning permission.

[135] Accordingly, I would respectfully disagree with the judge's conclusion (in [109] of the judgment) that the appellant be debarred from relying upon the argument based on affordable housing.

[136] In so far as Mr Drabble submitted that the appellant has no standing to bring the proceedings at all, I have no hesitation in rejecting that submission. The appellant has lived in Longstanton for upwards of 30 years. She plainly has a real and genuine interest in seeking to prevent the substantial development permitted by the planning permission.

[137] However, for the reasons I have given, I would dismiss this appeal.

LAWS LJ:

[138] I agree

ALDOUS LJ:

[139] I also agree.

Appeal dismissed.

Earthline Ltd *v* Secretary of State for Transport, Local Government and the Regions and another

[2002] EWCA Civ 1599

COURT OF APPEAL
Brooke and Keene LJJ and Bodey J
6 November 2002

A **Old mining permission – Planning and Compensation Act 1991 – Council condition imposing time limit on gravel extraction – Owner failing to challenge validity of new condition – Whether subsequent owner entitled to so challenge – Whether section 284 of Town and Country Planning Act 1990 precluding challenge to decision of Secretary of State**

B In 1946, the second defendant council, as the relevant mining authority, granted planning permission for gravel extraction on the subject site. This permission was subject to a mandatory time limit of 60 years, commencing February 1982, by virtue of section 7(5) of the Town and Country Planning (Minerals) Act 1981. Thus, the permission was to expire on 22 February
C 2042. In January 1992, the then landowner applied for the registration of the permission pursuant to section 22 of, and Schedule 2 to, the Planning and Compensation Act 1991. Under para 2(1)(c) of Schedule 2, a permission could not be made subject to a condition providing for the expiry of that permission before 22 February 2042. In 1996, in breach of the requirements of Schedule 2, the council made a determination that
D provided for the cessation of gravel extraction on or before September 2000 and for restoration works to be completed by September 2001. The then landowner did not challenge the council's determination. In 1999, the claimant, E, acquired the site and applied, under section 73 of the Town and Country Planning Act 1990, to insert 2002 and 2003 into the condition and sought a two-year extension. In January 2002, E appealed to the Secretary of State against the non-determination of that application. The
E inspector dismissed E's appeal. E's application to quash the inspector's decision and its contention that the council's 1996 decision to impose a time limit in breach of Schedule 2 was without effect were dismissed in the court below. The council appealed and E cross-appealed. The issues on appeal were whether: (i) E was entitled to mount a challenge to the validity of the condition when its predecessor in title did not appeal
F against the original determination; and (ii) the trial judge had been wrong in the exercise of his discretion to quash the inspector's decision on this ground.

Held The appeal was dismissed: the cross-appeal was allowed, and the decision remitted to the Secretary of State.

G The trial judge had been correct to hold that the language of the 1991 Act did not preclude a subsequent challenge to the validity of the condition and to be willing to decide the appeal in this way. The landowner had been given, by statute, an extension of its permission until 2042. Clear statutory language would be required in order to prevent the landowner's successors from challenging the council's misuse of power when they reduced the length of the permission so drastically: see [19]. E's cross-appeal was allowed and remitted to the Secretary of State, who would have to alter the condition to show that the permission was to extend to 2042: see [20]. A B

Cases referred to in the judgments

Boddington v *British Transport Police* [1999] 2 AC 143; [1998] 2 WLR 639; [1998] 2 All ER 203, HL

R v *North Yorkshire County Council, ex parte Brown* [2000] 1 AC 397; [1999] 2 WLR 452; [1999] 1 All ER 969; [1999] 1 PLR 116; [1999] JPL 116; [1999] JPL 616, HL C

R v *Wicks* [1998] AC 92; [1997] 2 WLR 876; [1997] 2 All ER 801; [1997] 2 PLR 97; [1997] JPL 1049

Tarmac Heavy Building Materials UK Ltd v *Secretary of State for the Environment, Transport and the Regions* (2000) 79 P&CR 260

D

Appeal against a decision of Sullivan J

This was an appeal by the second defendants, West Berkshire Council and a cross-appeal by the claimant, Earthline Ltd, against a decision by Sullivan J, allowing Earthline Ltd's first ground of appeal from an order of a planning inspector, dated 24 July 2001, dismissing its appeal against the council's failure to determine its application for the relaxation of a condition on an old mining permission. Sullivan J dismissed Earthline's second ground of appeal and Earthline cross-appealed. E

David Holgate QC and David Forsdick (instructed by Veale Wasbrough, of Bristol) represented the claimant, Earthline Ltd.

The first defendant, the Secretary of State for Transport, Local Government and the Regions, did not appear and was not represented. F

Timothy Straker QC and Andrew Tabachnik (instructed by Lester Aldridge, of Southampton) appeared for the second defendants, West Berkshire Council.

The following judgments were delivered. G

BROOKE LJ:

[1] This is an appeal by the second defendants, West Berkshire Council (the council), which are the local mineral planning authority, and a cross-appeal by the claimant, Earthline Ltd (Earthline), from an order by Sullivan J, allowing Earthline's appeal on its first ground of appeal from an order of a planning inspector, dated 24 July 2001, dismissing its appeal against the council's failure to determine its application for the relaxation of a condition on an old mining permission relating to land at Craven Keep, Hamstead Marshall. Sullivan J rejected Earthline's second ground of appeal. Hence its cross-appeal. The Secretary of State, although a party to H

A the proceedings before Sullivan J, did not appear to support the council's contentions on their appeal to this court.

[2] The history of this matter starts in 1946. On 18 June 1946, Newbury Rural District Council granted permission on an application under an interim development order (IDO) for gravel extraction on the land in question, subject to three fairly straightforward conditions. By section 77 B of the Town and Country Planning Act 1947, planning permission was deemed to have been granted under Part III of that Act for the same development and subject to the same conditions. The subsequent consolidating legislation, enacted in 1962 and 1971, maintained that position: see, most recently, para 3 of Schedule 3 to the Planning (Consequential Provisions) Act 1990.

C [3] The site was worked over many years, with restoration taking place in the late 1970s and early 1980s. Thereafter, it was, for all practical purposes, dormant until the events with which we are concerned in this appeal. The effect of section 7(5) of the Town and Country Planning (Minerals) Act 1981, which enacted one of the recommendations of the 1976 report of the Stevens Committee on "planning control over mineral D workings", was to impose a mandatory time limit of 60 years, from 22 February 1982, on the original permission, so that it would expire on 22 February 2042.

[4] On 22 January 1992, the then owner of the site applied for the 1946 permission to be registered pursuant to section 22 of, and Schedule 2 to, the Planning and Compensation Act 1991 (the 1991 Act), which had E recently come into force. On 22 September 1992, the council duly issued a notice of registration. On 13 November 1995, application was made for the determination of the conditions to which the permission was to be made subject. On 12 January 1996, the council duly determined those conditions. For the most part, these conditions were concerned with the orderly conduct of work on the site. The only condition with which we are F concerned on this appeal is condition 1, which provided that:

> The extraction of gravel shall cease on or before 30th September 2000. All restoration operations, including the removal of all plant, buildings, hardstandings and access roads, but excluding after-care and drainage, shall be completed by 30th September 2001.

G There was no appeal against the council's determination.

[5] It is common ground on this appeal that the council had no power to limit the period of time permitted for extracting gravels on the site in this way. The time limit for the permission was 22 February 2042. Paragraph 2(1)(c) of Schedule 2 to the 1991 Act provided that the new conditions "must include a condition that the winning and working of H minerals... must cease not later than February 21, 2042". What is in issue is whether the fact that no appeal was lodged against the council's deviant determination (so that anyone searching the register now would see the condition that limited the length of time the site could be used for gravel extraction) means that the new owner of the site is not now entitled to challenge the validity of the 1996 determination.

[6] The remainder of the history can be stated quite briefly. In due A
course, Earthline acquired the site, and, on 28 October 1999, it applied, pursuant to section 73 of the Town and Country Planning Act 1990 (the 1990 Act), for the dates 2002 and 2003 to be inserted into the body of condition 1 instead of the dates 2000 and 2001. After the date by which the council should have determined this application had expired, and after the
time for extracting gravel, as stated in that condition, had expired, B
Earthline appealed to the Secretary of State, on 9 January 2001, against the non-determination.

[7] By his decision letter dated 24 July 2001, the inspector dismissed the appeal. The grounds upon which he dismissed it are relevant only to Earthline's cross-appeal. Earthline duly applied to the High Court for an
order quashing the inspector's decision pursuant to section 288 of the C
1990 Act, and on this application, which was heard by Sullivan J, it raised for the first time the issue that is at the centre of the council's appeal to this court, namely that the 1946 permission did not expire until 2042, and that the council's attempt to reduce the time permitted for gravel extraction on the site by imposing condition 1 in its 1996 determination had had no
effect. D

[8] The scheme of the 1991 legislation relating to old mining permissions has now been before the courts on a number of occasions. It is clearly described by Lord Hoffmann in *R* v *North Yorkshire County Council, ex parte Brown* [2000] 1 AC 397[1] at pp 402D-403F. For the purposes of the present appeal, it is necessary to consider only the procedure under the
1991 Act for applying for the determination of the conditions to be E
imposed upon a "dormant permission": see Lord Hoffmann at p403E. This procedure is set out in section 22 and Schedule 2. Section 22 provides, so far as is material, that:

> (2) An old mining permission shall, if an application under that Schedule [2] to determine the conditions to which the permission is to be subject is
> finally determined, have effect as from the formal determination as if granted F
> on the terms required to be registered.
>
> (3) If no such development has, at any time in the period of two years ending with May 1, 1991, been carried out to any substantial extent anywhere in, on or under the land to which an old mining permission relates, that permission shall not authorise any such
> development to be carried out at any time after the coming into force of this G
> section unless –
>
> (a) the permission has effect in accordance with subsection (2) above; and
>
> (b) the development is carried out after such an application is finally determined.

[9] The arrangements by which conditions for old mining permissions
are to be determined are set out in para 2 of Schedule 2. The original H
application, made pursuant to para 2(2) and (3), suggested a termination

[1] [1999] 1 PLR 116

A date of 21 February 2024, no doubt because of a misprint for 2042 in Mineral Guidance Note 9, which misled the applicant. The council determined the conditions pursuant to para 2(6). Although the conditions differed from the conditions set out in the application, the applicant did not appeal to the Secretary of State pursuant to para 5(2). In those circumstances, the application for the determination of conditions was B finally determined when the time for appealing against the determination expired: see para 10(2). At that stage, the requirements of para 3(2) became operative:

> Where an application to determine the conditions to which an old mining permission is to be subject is finally determined, the conditions must be entered in the appropriate part of [the] register.

C

The condition setting out a final date of 30 September 2000 was therefore entered on the register. Earthline then appeared on the scene and sought a two-year extension.

[10] Although the council maintained before the judge that they were entitled to limit the permission in the way that they did, that contention is D no longer pursued. What is in issue on the appeal is whether Earthline is now entitled to mount this challenge to the validity of the condition when its predecessor in title did not appeal against the original determination, and whether the judge was wrong in the exercise of his discretion to quash the inspector's decision on this ground.

[11] One feature of the statutory scheme contained in the 1991 Act E deserves immediate attention. Section 284 of the 1990 Act contains the familiar provision that "except as provided in Part XII of that Act the validity of certain actions on the part of the Secretary of State shall not be questioned in any legal proceedings whatsoever". Paragraph 9(3) of Schedule 2 to the 1991 Act adds a decision of the Secretary of State on an appeal under para 5 of the Schedule to the other acts mentioned in Section F 284. It follows that if the council's determination had been appealed to the Secretary of State, and the landowner did not then take the matter to the High Court under section 288 of the 1990 Act, the validity of the Secretary of State's decision on the appeal could not thereafter be the subject of any legal contest.

[12] Whatever might be the consequence of this rule, however, on an G occasion when a landowner challenged a determination of modern conditions on different grounds from those connected with the invalidity of a condition such as condition 1 (which the council had no power to impose), the *no certiorari* rule in section 284 does not in any event apply to an unappealed determination of a mineral planning authority. The "finality" of their determination of modern conditions would not usually H be robust enough in itself to resist a challenge to the validity of one of the conditions on the ground that the authority had no power to impose it, if that challenge was otherwise sustainable in the context of a judicial review.

[13] The judge divided his approach to this issue into two parts. First, he said that he could see no reason why Earthline should not have been able to raise the validity of the condition before the inspector,

since it would have been a highly material consideration. In those circumstances, there was no reason to prevent Earthline from advancing the challenge on its application to the High Court. He said this because there was no need for any findings of fact to be made – the invalidity of condition 1 was apparent on its face – and it was incumbent upon the Secretary of State, when determining appeals, to direct himself correctly as to the applicable law. As I understand it, Mr Timothy Straker QC, who appears for the council, does not challenge the judge's willingness to entertain the point on arid procedural grounds. His substantive complaint is that the judge ought to have ruled that it was far too late for a challenge on the validity of the conditions to be launched.

[14] This, therefore, leads on to the judge's ruling on the substantive merits of the issue, which he expressed in these terms:

> I accept that the court has a discretion under section 288. It is not bound to quash a decision even if it concludes that there has been some error of law. The error of law in the present case went to the heart of the decision. Condition 1 was the only condition in dispute. It was vital that its invalidity was acknowledged as the correct starting point for any decision. In some cases the error of law may have become of academic interest, by reason, for example, of the passage of time. That is not the position here. As [Counsel] correctly submitted, this is an issue that will not go away. The validity of Condition 1 can be raised if the development commences and enforcement proceedings are begun, or if there is a prosecution for failure to comply with a breach of condition notice. It can be raised in the context of an application for a certificate under section 192 of the 1990 Act, or in proceedings for declaratory relief.
>
> In the light of these potential challenges, it seems eminently sensible to resolve the matter in the context of this decision letter.

[15] Although he was disposed to challenge the actual decision of the Divisional Court in one of the cases cited to us, Mr Straker did not really attempt to challenge the general propositions of law to which the judge referred. They were conveniently set out in the skeleton argument we received from Mr David Holgate QC, who appeared for Earthline, in these terms:

> Under general planning law invalid conditions may be challenged by a number of means:
>
> (1) In an appeal under section 174 against an enforcement notice alleging a breach of condition (see eg *Tarmac Heavy Building Materials UK Ltd* v *Secretary of State* (2000) 79 P&CR 260, 267-8;
>
> (2) As a defence to a prosecution for failing to comply with a breach of condition notice (*Dilieto* v *Ealing LBC* [2000] QB 381; *Davenport* v *Hammersmith and Fulham LBC* (1999) 78 P&CR 421;
>
> (3) By an application under section 192 of TCPA 1990 for a certificate as to the lawfulness of proposed operations;
>
> (4) Proceedings in the High Court seeking a declaration that condition 1 is invalid and that the 1946 permission will last until 2042 (*Hall & Co Ltd* v *Shoreham* UDC [1964] 1 WLR 240; *Mouchel Superannuation Fund Trustees* v *Oxfordshire* CC [1992] 1 PLR 97). Such well-established authorities have not

A been overruled by the comments of Lord Hoffmann in *R* v *East Sussex CC, ex parte Reprotech (Pebsham) Ltd* [2002] UKHL 8, [2002] 4 All ER 58.

[16] As to the first of these considerations, it should be observed that the effect of Jackson J's judgment in July 1999 in *Tarmac Heavy Building Materials UK Ltd* v *Secretary of State for the Environment, Transport and the Regions* (2000) 79 P&CR 260 was that a condition in a 1952 planning
B permission (to the effect that a concrete batching plant should be removed from a site once mineral working was completed) was declared invalid and unenforceable 47 years later.

[17] Mr Straker argued manfully that this particular statutory scheme precluded a challenge of this kind. He relied heavily upon the registration
C arrangements contained in the statutory scheme under the 1991 Act, and upon what Lord Hoffmann said in *Brown* about the significance of the register as a means of alerting the public to the existence of old mining permissions.

[18] In *R* v *Wicks* [1998] AC 92[1] the House of Lords made it clear that, on some occasions, a party to a planning matter might be
D precluded from challenging the validity of an action of a local planning authority even when it faced criminal sanctions. It all depended upon the proper interpretation of the relevant statute. This principle was reiterated by Lord Irvine of Lairg LC in *Boddington* v *British Transport Police* [1999] 2 AC 143, where he explained the decision in *Wicks* by reference to the elaborate code contained in the
E 1990 Act for appeals against enforcement notices. The existence of this code meant that it was inappropriate on a prosecution for breach of an enforcement notice for magistrates to have to concern themselves with difficult points of law that could have been taken either on an appeal against the notice or, in a rare case, on an application for judicial review.

F [19] In my judgment, however, the judge was right to hold that the language of the 1991 Act did not preclude a subsequent challenge to the validity of condition 1, and to be willing to decide the appeal in this way. After all, the landowner had been given, by statute, an extension of its permission to 2042, and it would need very clear statutory language to prevent its successors from challenging the
G council's misuse of power when they reduced the length of the permission so drastically. There was no evidence before us that any member of the public had actually been misled by seeing condition 1 on the register. If, in another case, evidence of this kind was forthcoming, a court would have to assess its significance and determine whether to grant a remedy to the landowner in the way that is now familiar
H on judicial review applications. If Mr Straker's submissions had been upheld in this case, the result would have had very wide

[1] [1997] 2 PLR 97

ramifications in the planning field outside the fairly narrow world of old mining permissions.

[20] The effect of upholding the judge's decision will be that Earthline's appeal will have to be remitted to the Secretary of State, who will have the power to alter condition 1 so that it shows that the permission extends to 2042. It is not apparent to me that any of the other conditions would have to be changed, but if the council argue that they should, then appropriate adjustments may be made on the remitted appeal.

[21] For these reasons, I would dismiss this appeal. It was common ground that if the appeal were dismissed, there would be no need for us to consider or make any order on the cross-appeal.

KEENE LJ:

[22] I agree. However comprehensive a code is achieved by the 1990 Act and its associated legislation, the fact remains that there is no equivalent to section 284 of that Act when one is dealing with a decision of a local planning authority, or, as in this case, a mineral planning authority. It would take clear and express language to remove from the courts their power to strike down an *ultra vires* condition, and such language does not exist in these circumstances. The House of Lords decision in *Wicks* was not concerned with whether the courts were prevented altogether, or after a certain date, from dealing with an allegation of unlawfulness, but only with the type of proceedings in which the issue could be raised.

[23] Once it is established that the court has the power to intervene, the issue becomes merely one of discretion. That discretion relates only to the question of remedy. In the present case, it is not disputed that the condition imposed was an unlawful one. In principle, an unlawful condition should be quashed, and nothing before us provides any proper basis for refusing to quash in this case. I too would dismiss this appeal.

BODEY J:

[24] I agree.

Appeal dismissed and cross-appeal allowed.

Withers *v* Secretary of State for Transport, Local Government and the Regions and another

[2002] EWHC 2425 (Admin)

QUEEN'S BENCH DIVISION: ADMINISTRATIVE COURT
Sullivan J
11 November 2002

A **Curtilage – Town and Country Planning Act 1990 – Annex C to PPG 3 – Whether stables forming part of curtilage – Whether inspector misapplying PPG 3 – Whether special circumstances outweighing the presumption against development within green belt**

The second defendant council refused the claimant's application
B for planning permission to redevelop, for residential use, a former livery stable situated within a green belt area. The Secretary of State, by his inspector, dismissed an appeal against that decision, and concluded that the proposal constituted inappropriate development within the green belt and that there were no special circumstances outweighing the presumption against such development. The claimant challenged the
C inspector's decision that the appeal site represented "previously developed land with urban uses within an urban area". He argued that, in light of the definition in footnote 2 to Annex C to PPG 3, the inspector should have considered the extent to which the open land around the stable buildings fell within the curtilage of the stables, and therefore within the definition of previously developed land for the purposes of
D Annex C. The claimant contended that if the inspector had reached such a conclusion, it might have made a difference to his consideration of the question of special circumstances.

Held The application was dismissed.

The inspector was entitled to reach the conclusion he had. There was no basis for concluding that the appeal site constituted previously
E developed land. With regard to the contention that the inspector should have considered whether the paddocks formed part of the curtilage of the stables, it should be borne in mind that the advice in PPG 3 has to be applied in a common-sense fashion. The inspector was not interpreting a statutory instrument, but applying policy guidance: see [9-10]. Even if it can sensibly be said that grazing paddocks are within
F the curtilage of a stable, the footnote to Annex C of PPG 3 makes it perfectly clear that only that part of a curtilage containing building footprints should be redeveloped as previously developed land in accordance with the advice in Annex C: see [12]. The question for the inspector in the instant case concerned the extent of the existing buildings and the open area, and whether these uses could sensibly be regarded as
G urban uses in an urban area or were really the kind of rural use that one might expect to find in the green belt. Whether the entire site should have

been regarded as previously developed land was not the critical factor in the inspector's determination. The critical factor was that, on the ground, the site was more akin to an open agricultural use than an urban development within a built-up area. That was a wholly unexceptional conclusion. It follows that there was no force in the sole criticism of the decision: see [14]. A

B

No cases were referred to in the judgment

Application under section 288 of the Town and Country Planning Act 1990

This was an application by the claimant, David Withers, under section 288 of the Town and Country Planning Act 1990, against a decision of the first defendant, the Secretary of State for Local Government, Transport and the Regions, dismissing an appeal against a refusal of planning permission by the second defendants, North Somerset Council. C

Robert White (instructed by Burroughs Day, of Bristol) appeared for the claimant, David Withers.

Philip Coppel (instructed by the Treasury Solicitor) represented the first defendant, the Secretary of State for Local Government, Transport and the Regions. D

The second defendants, North Somerset Council, did not appear and were not represented.

The following judgment was delivered. E

SULLIVAN J:

[1] This is a challenge, under section 288 of the Town and Country Planning Act 1990, to an inspector's decision dismissing an appeal against a refusal of planning permission by the second defendants for the redevelopment of former livery stables for residential use at South View Farm, 82 Clevedon Road, Tickenham, Clevedon, North Somerset. F

[2] The appeal was determined by written representations. The inspector made a site visit on 9 April, and his decision dismissing the appeal is contained in a decision letter dated 2 May. The inspector described the nature of the proposal in para 1 of the decision letter: G

> the proposal would involve the demolition of the existing buildings, including livery stables, tack rooms and kennels, and the redevelopment of this 0.8ha site for new housing.

[3] The site is within the South West Avon green belt. Thus, the inspector identified the main issues as being whether the proposal represented appropriate development in the green belt and, if not, whether there were any very special circumstances that would outweigh the presumption against inappropriate development in such areas. The inspector concluded that: (a) the proposal was not appropriate development in the green belt; and (b) there were no very special H

circumstances that outweighed the presumption against inappropriate development in the green belt.

[4] On behalf of the claimant, Mr Robert White does not challenge the inspector's conclusion (a), that is to say that the proposal was indeed inappropriate development in the green belt. He challenges the inspector's conclusion (b), that is to say the finding that there were no very special circumstances to justify inappropriate development.

[5] The alleged very special circumstance advanced by the claimant in his written representations was that the appeal site represents:

> *previously developed land* with *urban* uses within an *urban area.*

(See para 11 of the decision letter.) In that paragraph, the inspector referred to Annex C of PPG 3, and said that it defined previously developed land as land:

> *"which is or was occupied by a permanent structure and associated infrastructure."* It confirms that the definition covers the whole curtilage of the site and acknowledges that such land can occur in both built-up and rural settings. However, it excludes agricultural and forestry buildings and land and buildings that are currently in agricultural or forestry use.

In para 12, the inspector went on to say:

> Since the appeal site contains existing buildings, built of concrete blocks with asbestos roofs, with a driveway and surfaced areas, there can be little dispute that at least part of the site has been previously developed. I understand that the existing buildings were previously used as commercial livery stables and tack rooms, rated for commercial purposes, rather than for agriculture. Under the *"Land Use Changes in England"* classification, such a use would fall under definition of *Leisure & Recreational Buildings* associated with outdoor recreation within the *urban* category of land uses. I also note that the *MAFF* survey indicates that part of the appeal site is in non-agricultural *urban* use.

Pausing there, Mr White submits that the inspector fell into error in concluding that part of the site, as opposed to the whole site, had been previously developed. Previously developed land, as defined in Annex C to PPG 3:

> is that which is or was occupied by a permanent structure (excluding agricultural or forestry buildings), and associated fixed surface infrastructure. The definition covers the curtilage of the development. Previously developed land may occur in both built-up and rural settings.

There are a number of footnotes, and footnote 2 is as follows:

> The curtilage is defined as the area of land attached to a building. All of the land within the curtilage of the site (as defined above) will also be defined as previously developed. However, this does not mean that the whole area of the curtilage should therefore be redeveloped. For example, where the footprint of a building only occupies a proportion of a site of which the remainder is open land (such as at an airfield or a hospital) the whole site should not normally be developed to the boundary of the curtilage. The local

> planning authority should make a judgment about site layout in this context, bearing in mind other planning considerations, such as policies for the protection of open space and playing fields or development in the countryside, how the site relates to the surrounding area, and requirements for on-site open space, buffer strips, landscaped areas etc.

[6] It is submitted on behalf of the claimant that, in the light of that definition, the inspector should have considered the extent to which the open land around the stable buildings fell within the curtilage of the stables, and thus within the definition of "previously developed" land for the purposes of Annex C. It is further submitted that had the inspector considered that matter, and had he concluded that all the site fell within the definition of previously developed land because it was all within the curtilage of the stables, then that might have made a difference to his consideration of the question of special circumstances.

[7] Whatever view might be formed as to the first of those propositions, there can be no doubt, in my judgment, that the second proposition is wholly unfounded once the remainder of the decision letter is considered. The inspector said in para 13:

> 13. However, it does not necessarily follow that the appeal site is predominantly in *urban* use within an *urban* area. Apart from the existing buildings, the remainder of the site is open and largely undeveloped. It has been used for the grazing and exercising of horses and livestock, which could not reasonably be thought of as an *urban* use. In my view, it is more closely related to outdoor recreation and agriculture which are defined as appropriate uses for Green Belt land. Nor is it *derelict* urban land which might assist urban regeneration. Furthermore, the site lies outside the settlement boundary of Tickenham which is a rural settlement in the Green Belt. It is separated from the main built-up area of the village by paddocks and the main road and lies on the exposed southern fringe of the settlement, overlooking Tickenham Moor. It is neither within nor adjoining an *urban* area. Consequently, it cannot reasonably be concluded that this is an *urban* site with predominantly *urban* uses, even though it may have been used for purposes which are considered *urban* in general land-use classification terms. In my view, the site partly comprises *previously developed land* within a rural area, with uses more akin to open agricultural uses than those more typical of urban development within a built-up area.

Pausing there, if one replaces the words "partly comprises previously developed land" in the last sentence of para 13 with "comprises previously developed land" one finds: "the site comprises previously developed land within a rural area, with uses more akin to open agricultural uses than those more typical of urban development within a built-up area". If that is done, it can be seen that the contention that there might be special circumstances does not begin to get off the ground whatever view is taken of the curtilage argument.

[8] If one goes on to para 14, the inspector said:

> In any event, PPG 3 confirms that even though part of the site may comprise previously developed land, this does not alter the general presumption against inappropriate development in the Green Belt. Nor does it change the

> policy in respect of *major* sites within the Green Belt in PPG 2 (Annex C) (even though the appeal site would not fall into this category.) Consequently, while the fact that part of the appeal site constitutes *previously developed land* is a material planning consideration in favour of its development, it does not, in my view, outweigh the general presumption against inappropriate development in the Green Belt or constitute the very special circumstances needed to justify new residential development in the Green Belt.

[9] Again, I can see no basis whatsoever for the proposition that if the inspector had concluded that the whole of the appeal site constituted previously developed land, the conclusion in para 14 might have been expected to be different. The inspector concluded that the "previously developed land" argument did not amount to a very special circumstance, not because only part of the site fell within that definition, but because there was no suggestion in PPG 3 that this might amount to a very special circumstance. The inspector was saying, in short, that, as a matter of fact: there were a number of buildings on this land, which was predominantly open; the buildings and uses were akin to open agricultural uses typical of a rural area; and the fact that they were there did not amount to a very special circumstance justifying residential development in the green belt. All those conclusions hold good whatever view one takes of the definition of "previously developed land" within Annex C to PPG 3.

[10] Reverting, therefore, to the first proposition, that is to say, that the inspector should have considered whether the paddocks formed part of the curtilage of the stables, it has to be borne in mind that the advice in PPG 3 must be applied in a common-sense fashion. The inspector was not interpreting a statutory instrument; he was applying policy guidance.

[11] Plainly, if one is considering a green belt case, the extent to which a particular site is open, as opposed to covered with buildings, is a highly material consideration. Thus, the inspector was perfectly entitled to look at the situation on the ground, and to say that, apart from the existing buildings, the remainder of the site was open and largely undeveloped; it had been used for grazing and the exercising of horses, precisely the kind of use that one would expect to find in a rural area.

[12] Thus, the conclusion in para 13 was an eminently sensible one. It would not have been falsified by a consideration of the curtilage point. It has to be borne in mind that even if it can sensibly be said that grazing paddocks are within the curtilage of a stable, the footnote makes it perfectly clear that it is only that part of a curtilage where there are building footprints that should be redeveloped, in accordance with the advice in Annex C, as previously developed land. Even if the inspector had concluded that the paddocks were within the curtilage of the stables, he would immediately have reminded himself that it did not follow that they should be regarded as being developable, given the remaining text of the footnote.

[13] Thus, it seems to me that the sole criticism of the inspector's decision leads nowhere, and is purely academic.

[14] The question for the inspector in this green belt case was the extent of the existing buildings and the extent of the open area, and whether

these uses could sensibly be regarded as urban uses in an urban area, or really the kind of rural use that one might expect to find in the green belt. It is not submitted that there is any indication in PPG 3 that previously developed land should be regarded as an exception to green belt policy as set out in PPG 2. It is not suggested that this is, for example, a major site to which specific policies in PPG 2 are applicable. In short, all the inspector was saying was that a wish to redevelop former livery stables in the countryside for residential purposes did not amount to very special circumstances for the purposes of PPG2. Whether part or all of the site should have been regarded as previously developed land was not the critical factor in the inspector's determination. The critical factor was that, on the ground, the site was more akin to an open agricultural use than an urban development within a built-up area. That was a wholly unexceptional conclusion. It follows that there is no force in the sole criticism of this decision.

[15] For these reasons, this application must be refused.

Application dismissed.

Henry Boot Homes Ltd *v* Bassetlaw District Council

[2002] EWCA Civ 983

COURT OF APPEAL
Brooke and Keene LJJ and Bodey J
28 November 2002

A **Planning permission – Planning conditions – Conditions requiring steps to be carried out "before development commences" – Conditions stipulating "no development of the site shall begin" until specified steps carried out – Non-compliance with conditions – Works commencing within five-year planning permission validity period – Whether works amounting to valid commencement of development**
B **permitted by planning permission – Outline planning permission – Time limits – Whether planning permission implemented within time limits – Whether outline planning permission remaining extant – Whether developer having legitimate expectation that permission treated as implemented – Whether conditions waived**

C In August 1995, the respondent council granted outline planning permission to the appellant developer for the residential development of a large site, subject to conditions, including a condition requiring the approval of reserved matters. Reserved matters approval was given on 5 December 1995, subject to several conditions that stipulated either that something was to be done "before development commences" or that
D "no development of the site shall begin until" some action had been carried out. Works sufficient in physical terms to amount to a "material operation" within the meaning of section 56 of the Town and Country Planning Act 1990 began on 16 January 1996. However, they were commenced without compliance with the conditions of the outline planning permission, or the reserved matters approval that
E required some action prior to the commencement of development. Between 1996 and the commencement of the proceedings, there had been discussions between the parties with regard to the regularisation of the development, and it was not until 1999 that it was first suggested that the planning permission had never been implemented. In the court below, Sullivan J held that, as the commenced development was not
F authorised by the planning permission, and as no such authorised development had commenced prior to 2 August 2000, that being the expiry date of the five-year planning permission validity period, the planning permission was no longer extant. The appellant appealed, contending that it had a legitimate expectation, based upon representations made on behalf of the council, either that the permission
G would be treated as having been implemented, or that compliance with the conditions had been waived.

Held The appeal was dismissed.

The code of planning control provides the public with the right to be informed and to object. Accordingly, the scope for the variation or discharge of planning condition requirements by some other non-statutory method, thereby bypassing the statutory safeguards for the public, are extremely limited. The interests of third parties and the public also greatly reduce the potential for a legitimate expectation to arise. Any expectation by the appellant that its works would be treated as a lawful implementation was not a legitimate expectation. There was no lawful waiver of conditions.

Cases referred to in the judgment

Agecrest Ltd v *Gwynedd County Council* [1998] JPL 325

Coghurst Wood Leisure Park Ltd v *Secretary of State for Transport, Local Government and the Regions* [2002] EWHC 1091 (Admin); [2002] 24 EG 145 (CS), AC

Etheridge v *Secretary of State for the Environment* (1983) 48 P&CR 35; [1984] JPL 340, QB

FG Whitley & Sons Co Ltd v *Secretary of State for Wales* (1992) 64 P&CR 296; [1992] 3 PLR 72; [1992] JPL 856, CA

Henry Boot Homes Ltd v *Bassetlaw District Council* [2002] EWHC 546 (Admin); [2002] JPL 1224; [2002] 13 EG 100 (CS)

Leisure Great Britain plc v *Isle of Wight County Council* (2000) 80 P&CR 370

Oakimber Ltd v *Elmbridge Borough Council, sub nom R* v *Elmbridge Borough Council, ex parte Oakimber Ltd* [1991] 3 PLR 35; (1991) 62 P&CR 594; [1992] JPL 48

Pioneer Aggregates (UK) Ltd v *Secretary of State for the Environment* [1985] AC 132; [1984] 3 WLR 32; [1984] 2 All ER 358; (1984) 82 LGR 488; 48 P&CR 95; [1984] 2 EGLR 183; 272 EG 425; [1984] JPL 651, HL

Powergen UK plc v *Leicester City Council; sub nom R* v *Leicester City Council, ex parte Powergen UK plc* (2001) 81 P&CR 5; [2000] JPL 1037, CA

Powergen UK plc v *Leicester City Council; sub nom R* v *Leicester City Council, ex parte Powergen UK plc* (2000) 80 P&CR 176; [1999] 4 PLR 91; [2000] JPL 629, QB

R (on the application of Reprotech (Pebsham) Ltd) v *East Sussex County Council* [2002] UKHL 8; [2002] 4 All ER 58; [2002] 2 PLR 60; [2002] JPL 821

R v *Bassetlaw District Council, ex parte Oxby* [1998] PLCR 283

R v *Flintshire County Council, ex parte Somerfield Stores Ltd* [1998] PLCR 336, QBD

R v *Inland Revenue Commissioners, ex parte MFK Underwriting Agencies* [1990] 1 WLR 1545; [1990] 1 All ER 91

R v *Inland Revenue Commissioners, ex parte Unilever plc* [1996] STC 681, CA

R v *North and East Devon Health Authority, ex parte Coughlan* [2001] QB 213; [2000] 2 WLR 622; [2000] 3 All ER 850; [1999] Lloyd's Rep Med 306, CA

R v *Secretary of State for the Environment, ex parte Percy Bilton Industrial Properties Ltd* (1975) 74 LGR 144; 31 P&CR 154

Tesco Stores Ltd v *North Norfolk District Council* (1999) 78 P&CR 359; [1999] EG 39 (CS); [1999] JPL 920, CA

Appeal against a decision of Sullivan J

This was an appeal by Henry Boot Homes Ltd from a decision of Sullivan J, who had dismissed an application by the appellant for a declaration as to the validity of a planning permission issued by the respondents, Bassetlaw District Council.

David Holgate QC and Timothy Morshead (instructed by Hammonds Suddards Edge, of Leeds) appeared for the appellant, Henry Boot Homes Ltd.

Mark Lowe QC and Jonathan Clay (instructed by the solicitor to Bassetlaw District Council) represented the respondent council.

The following judgment was delivered.

KEENE LJ:

Introduction

[1] The Town and Country Planning Act 1990 (the Act) makes provision for the expiry of planning permissions not implemented within a certain time. It does this by requiring any planning permission that is granted to be subject to conditions under which the development to which the permission relates must be begun within a certain period.

[2] If the development has not been begun by the end of that period, it is no longer authorised by the permission: section 93(4)(a). The Act also defines when it is that development begins. Section 56(2) provides that:

> development shall be taken to be begun on the earliest date on which any material operation comprised in the development begins to be carried out.

"Material operation" is then defined so as to include various kinds of works, including any work of construction in the course of the erection of a building: section 56(4).

[3] The conditions imposing the time limits are to be found in sections 91 and 92 of the Act, the latter dealing with outline planning permissions as opposed to full planning permissions. The present appeal is concerned with an outline planning permission, but the issues raised affect all planning permissions covered by either of those two sections. It is unnecessary to set out the terms of those two sections in this judgment because the planning permission involved in the present case contained conditions complying with the Act; conditions to which I shall refer in due course. In addition to those time limit conditions, planning permissions regularly include a number of other conditions relating to the development. This appeal concerns whether, and in what circumstances, development under a planning permission can be said to have begun as a result of a material operation carried out in breach of a condition on the permission, and to what extent there can be a legitimate expectation on the part of a developer that the local planning authority will treat the material operation as authorised by the permission despite the breach of condition.

Facts

[4] The facts of this case were set out in detail by Sullivan J in his judgment in the court below, a judgment reported at [2002] EWHC 546 (Admin); [2002] JPL 1224. However, it is said by the appellant, Henry Boot Homes Ltd, that that judgment does not refer to all the material facts, and so it is necessary to cover them afresh in this judgment, particularly

because of the emphasis being placed by the appellant upon a legitimate expectation resulting from the respondents' conduct.

[5] The case concerns a site of some 13ha (32 acres), or thereabouts, located in the valley of the River Idle at Retford, Nottinghamshire. The respondents are the local planning authority for the area. The river forms the eastern boundary of the site. To the west lies residential development served by a number of estate roads, such as Heathfield Gardens and Camborne Crescent, leading into Hallcroft Road. To the north there is open land, as there is to the immediate south, with a retail store and a road called Amcott Way beyond it.

[6] The site can be regarded as being divided into two parts by a drainage ditch running roughly north to south. The smaller western part is referred to in the documents as part A or site A. On part of this area, some 63 houses have now been built, with part of an estate road having been laid out on the remainder. The larger eastern part, B, is undeveloped and some of it is part of the floodplain of the River Idle.

[7] In May 1994, at which time the whole site was undeveloped, outline planning permission for residential development of the whole site was granted, contrary to the recommendation of officers and despite objection from the highway authority, the National Rivers Authority, and some local residents. The only vehicular access obtainable as things stood was via the residential estate road, Heathfield Gardens, and onto Hallcroft Road, where there were already traffic problems. The permission, which envisaged some 315 dwellings, was subject to a number of conditions, some of which required work to be done, or details to be approved, before any development began. It is unnecessary to set them out because they were incorporated into a later permission of greater significance for this appeal. But one of the conditions, condition 3, required that the road layout on site should include a particular feature, and that no dwelling should be commenced until that road feature had been provided.

[8] Later that year, the appellant bought the site, although it indicated that it had "serious reservations about the effect of the floodplain of the River Idle on the nett developable area". It then applied, in April 1995, for an amendment to condition 3 of the 1994 permission, so that the road layout could be varied in a particular way. This was an application under section 73 of the Act, which enables an application to be made:

> for planning permission for the development of land without complying with conditions subject to which a previous planning permission was granted. (section 73(1))

[9] On such an application, the local planning authority (LPA) can consider only the question of the conditions to be attached, but the end result, if it is decided that permission should be granted subject to different conditions from those in the original permission, is a wholly new planning permission. That is what happened here. On 2 August 1995, the respondents granted a new outline permission similar to the 1994 permission but with a different condition 3. This outline permission (the

A outline permission) is central to this case. It contained, first, the time-limit condition required by section 92(2) of the Act:

> 1. Application for approval of reserved matters must be made not later than the expiration of three years beginning with the date of this permission and the development must be begun not later than whichever is the later of the following dates:
>
> B (a) the expiration of five years from the date of this permission; or
>
> (b) the expiration of two years from the final approval of the reserved matters...

Condition 2 was the standard reserved matters condition:

> 2. The siting, design and external appearance of the building(s), the means C of access thereto, and the landscaping of the site shall be only as may be approved in writing by the District Planning Authority before any development commences.

Condition 3 was as follows:

> The details required by condition No 2 above shall provide for an adoptable D road layout which shall include an extension of Heathfield Gardens in a easterly direction, making provision for access to adjoining land to the south in the manner indicated in Drawing No BH6616/1F...
>
> No dwelling shall be commenced until the extension to Heathfield Gardens, including the roundabout and the access to the land to the south have been constructed to at least base course level, from the existing end of Heathfield Gardens to the point where it meets the southern boundary site.

E

One can then go to condition 8:

> No development of the site shall begin until such time as full details of the manner in which foul sewage and surface water are to be disposed of from the site have been submitted to and agreed in writing by the District Planning Authority.

F

Condition 10:

> The landscaping scheme required by Condition No 2 shall be submitted to and approved in writing by the District Planning Authority before development commences...

G [10] On 8 September 1995, the appellant applied for approval of the reserved matters under condition 2. Various discussions and meetings took place, including ones involving the National Rivers Authority, who were concerned about flooding. The appellant stated that it wanted to begin development as soon as possible and, according to the minutes of one meeting "agreed to limit the first phase to land outside the H floodplain".

[11] On 17 November 1995, even before the application for reserved matters approval had been determined, the appellant wrote to the respondents to confirm that development would commence in January 1996. An application was then made for a building regulation approval for a number of the proposed dwellings.

[12] On 5 December 1995 the reserved matters application was approved, subject to a number of conditions, including:

> 2. No dwelling shall be commenced until the extension of Heathfield Gardens has been constructed, and surfaced to at least base course level, from the existing end of Heathfield Gardens to the point where it meets the southern boundary of the site...
>
> 5. Before development commences precise details of the finished floor level of each dwelling, road and footpaths, garden areas and open spaces shall be submitted to and agreed in writing with the District Planning Authority...
>
> 6. The facing and roofing materials to be used in the development hereby permitted shall be only as may be agreed in writing by the District Planning Authority before development commences.
>
> 7. A scheme for tree planting on and landscaping treatment of the site, including the area indicated as Public Open Space, shall be submitted to and agreed in writing by the District Planning Authority before development commences...
>
> 8. The form of surfacing used for all outdoor hard surfaces on the site shall be only as may be agreed in writing by the District Planning Authority before development commences.
>
> 9. Precise details of the landscaped strip adjacent to the River Idle shall be submitted to and agreed in writing by the District Planning Authority before development commences...
>
> 10. No development of the site shall begin until such time as full details of the manner in which foul sewage and surface water are to be disposed of from the site have been submitted to and agreed in writing by the District Planning Authority.
>
> 12. Precise details of the landscaping, surfacing treatment and footpath provision for the strip of land containing the gas main shall be submitted to and agreed in writing with the District Planning Authority before development commences...

[13] There is no dispute that this approval (the detailed approval) met condition 2 of the outline permission dealing with those matters referred to as "reserved matters", and, consequently, the three-year time limit for approval of reserved matters, set out in condition 1 of the outline permission, was met. All that remained was to begin the development within five years of the date of the outline permission, that is to say by 2 August 2000. Likewise, there is no dispute about two further facts: first, works sufficient in physical terms to amount to a "material operation" within the meaning of section 56 of the Act began on 16 January 1996 and continued for a considerable time thereafter, as I shall describe; second, as Sullivan J noted, such works began without compliance with condition 8 of the outline permission (covered also by condition 10 of the detailed consent) and without compliance with conditions 5, 6, 7, 8, 9, 10 and 12 of the detailed consent. Each of those conditions stipulated either that something was to be done "before development commences" or that "no development of the site shall begin until" some action had been carried out.

A [14] There had been an application on 8 December 1995 by the appellant for approval of an external finishing schedule, in effect under condition 6 of the detailed approval, for what was described as "the first phase" of the development. This was approved by the respondents on 2 February 1996, but the letter of approval went on to say:

> B I would also take this opportunity to highlight other conditions of [the] planning application which have yet to be complied with, I trust these matters will be given attention at the earliest opportunity.

There was likewise an approval on 24 April to landscaping proposals for part of the site, described as "phase III", so that there was partial compliance with condition 7 on the detailed approval. Once again, the
C officer writing on behalf of the respondents reminded the appellant of the need for "full compliance" with that condition and of the requirements of the other conditions.

[15] However, in the meantime, other more dramatic events had been taking place. During 1995, concern had been growing about the number of planning permissions that had been granted by
D the respondent authority against professional advice, almost always to the benefit of a particular individual developer. This individual had been the driving force behind the company that had obtained the 1994 outline permission on the site with which we are concerned, and the 1994 permission was one of those causing concern. The result was the appointment by the full council of the respondents, in November 1995,
E of Mr Richard Phelps CBE to carry out an independent investigation of those concerns. He reported to the respondents in March 1996. His report was highly critical of the respondents' planning committee. Paragraph 2(xii) stated:

> A reasonable outside observer must conclude that the Committee has been
> F manipulated in some way. Otherwise there is no rational explanation to the way in which the Hallcroft Road decision was reached, or the outline planning permission for site 9 at Ordsall.

The Hallcroft Road decision was the grant, in 1994, of permission for the site in this appeal. The outline planning permission for the site at Ordsall was subsequently quashed in judicial review proceedings brought by the
G leader of the council, Mr Oxby: see *R* v *Bassetlaw District Council, ex parte Oxby* [1998] PLCR 283.

Mr Phelps recommended that "to restore confidence in the planning system in *Bassetlaw*, the present planning committee should resign en bloc as soon as practicable".

H His report also stated:

> The Council should decide within about 6 months whether in relation to the major part of the Henry Boot site in the Idle Valley (ie the "212 houses"), they should revoke the planning approval completely or modify it and/or meet the cost of the construction of a connecting road to Amcott Way.

The "major part of the Henry Boot site" there referred to is the area known as part B. It is clear that it was that part that gave rise to the significant planning objections. A

[16] I should say, at this point, that there is no criticism of the appellant company in the Phelps Report, and it is accepted by the respondents that the appellant bought the site in good faith. It is, as Sullivan J commented, a reputable housebuilding company. None the less, upon the receipt of the B Phelps Report, the respondents had to consider what was to be done about this site or at least the major part of it. The report had emphasised the severity of the damage that would be done to the community were part B of the site to be developed for housing.

[17] On 2 May 1996, the respondents wrote to the appellant, referring to Mr Phelps' recommendation that the authority should decide "within C about 6 months" whether to revoke or modify the permission in respect of part B of the site and/or to meet the cost of constructing a road running south to Amcott Way. The letter continued:

> Council officers have begun to consider available options in relation to this matter and will wish to discuss these with you in the near future. D
>
> However, in the meantime, I am writing to you specifically in relation to the latest approval granted on 5th December 1995. There are a number of conditions of the approval which have not been complied with notably 2, 3, 5, 8, 9, 10 and 12. Clearly consideration needs to be given to providing appropriate details to deal with the requirements of these conditions. My purpose in writing is in relation to condition 2.
>
> I believe it is unlikely that the Council will seek to revoke or modify the E permission in so far as it relates to that part of the site to the west of the dyke [Site A]. However, one of the options available to the Council is to consider revocation or modification of the permission in relation to the remainder of the land. In these circumstances, I would suggest that full compliance with the requirements of condition 2 would be inadvisable and I can confirm that no action to ensure such compliance will be taken by the Council at this time.
>
> An appraisal of the land in the Idle Valley has been started and other F information is being sought from the highways authority and the Environment Agency. The intention is to reach conclusions in this matter as soon as possible. Obviously, your firm will need to be involved in formulating decisions on this.

Condition 2 on the detailed approval was the one that required the G construction and surfacing to base-course level of a road from Heathfield Gardens to the southern boundary of the site.

[18] There was then a meeting between representatives of both parties to discuss the options, with notes being taken of the discussion. Those reveal that representatives of the appellant had expressed concern that the council were viewing their site in two parts; they had bought it as a whole, H designed it as a whole and any phasing was set for the convenience of the developer.

It was also recorded that, as between revocation and modification of the permission, the council's view was that modification was more likely since Henry Boot had already started to develop part of the site.

A This meeting was subsequently referred to in a letter from the appellant to the respondents' head of legal services, dated 4 July 1996, in which it was said:

> As you acknowledged, the development has commenced and the permission has been implemented. The company is presently investing heavily in the site with infrastructure works, roads and sewers, show homes and
> B production units.

In August, a local resident wrote to the council saying that houses were being constructed and was told that the development was being carried out under the 1995 outline permission.

At a meeting between representatives of both parties in that same
C month, the chief executive of the respondents, Mr Havenhand, referred to the judicial review proceedings being taken in respect of the planning permission granted for the site at Ordsall. The appellant's minute of this meeting records that:

> Mr Havenhand reiterated that the current objective was to establish that Ordsall was an unlawful decision and the real aim of the judicial review is to
> D provide substantive proof to the police. Mr Havenhand went on to acknowledge our planning permission was valid.

[19] In November 1996, the respondents gave written approval for details of floor levels, roads and footpaths, gardens and open areas for the westernmost part of the site under condition 5 of the detailed approval.
E The letter noted that the details:

> discharge the requirements of Condition No 5 for this part of the site. The condition will only be fully discharged when the necessary levels details for the whole of the site have been agreed in writing.

An internal note by an officer of the respondents, undated but probably in
F about November 1996, observed that the appellant was in breach of conditions, but the writer thought that matters could be resolved simply. The note made reference to "the first phase" of the development. This and other documented material shows that, by this time, the respondents' officers were aware that the appellant had started work on the site and had done so in breach of condition. For example, in February 1997, a
G memorandum from the head of planning services of the council notes that the appellant had failed to comply with the terms of the planning consent, in that it had not supplied information on floor levels etc, required by that consent, before building commenced.

[20] There was then a meeting between representatives of the appellant and respondents on 24 February 1997 in an attempt to arrive at a solution
H to the problems concerning the site in the light of the Phelps Report. The appellant was expressing concern at the delays. The respondents proposed that access to part B of the site should be obtained other than via Heathfield Gardens, which would then serve only what was referred to as the first phase. The appellant objected that the council were attempting to phase the development by artificial means.

[21] None the less, discussions about a possible compromise went on, A so that, by early May 1997, the officers were recommending that the council should look to find solutions rather than opting for revocation of the 1995 outline permission. Further studies were undertaken, and, eventually, on 22 December 1997, there was a report by officers to a special council meeting. This set out the various options, including revocation of the permission in so far as it related to part B of the site, purchase of that B part of the site by the council, or an alternative form of development on that area. The last of those options would have reduced development in the flood plain and amended the road layout. It required a new planning application and permission. It was this last option that found favour with the council, and it was therefore contemplated that the appellant would submit a new planning application. C

[22] The respondents' officers reminded the appellant again in July 1998 that it had not yet complied with most of the conditions of the existing consent. In that same month, the respondents published a number of topic papers for the forthcoming local plan inquiry, including one dealing with the Phelps Report. In the course of this paper, it was said that a valid planning permission existed on the appellant's site, and that D implementation of the permission had begun. Some of the houses were occupied. Similar comments were subsequently made in proofs of evidence used by the respondents at the local plan inquiry.

[23] On 7 August 1998 and 11 September 1998, the appellant submitted its revised scheme, in the shape of applications for planning permission for housing on the undeveloped parts of the site, namely the balance of E part A and the whole of part B. The local plan inquiry itself began in September 1998, and heard objections from a number of parties to the proposed allocation of this site for housing development. One of those objectors was another housebuilder, Aldergate Properties Ltd (Aldergate). Subsequently, Aldergate wrote to the respondents threatening judicial review if they dealt with the appellant's pending planning applications F before the local plan inspector had reported. The appellant expressed concern about the delay, but the respondents decided that they had to take counsel's opinion on the matter.

[24] Counsel advised in writing on 12 February 1999. He had been instructed: "Development within part A has already commenced pursuant to the extant permission and some 40 dwellings have already G been constructed." Much of his opinion was devoted to the issue of the alleged prematurity of any decision by the respondents on the appellant's applications, but he also addressed the question of the weight to be given to the existing planning permission. He referred to doubts as to whether that permission would in fact be implemented in full, and then stated: H

> If, as with the access pre-conditions, certain works cannot or will not be implemented, then all development is taking place in breach of planning control. If those works are not practicably capable of being implemented then the very validity of the permission itself may be in question.

A [25] This seems to have been the first time that any question as to the validity of the outline permission had been raised. Counsel advised that an assessment of the viability of implementation of that permission be made, and the respondents' planning committee accepted that recommendation on 3 March 1999.

[26] Following reminders to the appellant that it needed to comply with B the conditions on the permission, the respondents' head of legal and estate services, Mrs Brown, wrote, on 14 May 1999, as follows:

> I note from our meeting that your Company probably intended to recommence building in advance of the Council's consideration of the outstanding planning application for the Site.
>
> C You will be aware that the permission is subject to a number of conditions that have yet to be complied with. In view of the sensitivity of the Site the Council would expect that you comply with all of these conditions before you commence any further works. In particular further information must be provided on the proposed levels and landscaping for the site as a whole in compliance with Conditions 5, 9 and 12.

D On that same day, she sought further advice from counsel, noting that the appellant was in breach of several conditions on the permission and asking for his views on the expediency of issuing an enforcement notice. She also raised the issue of whether the validity of the permission could be challenged because of the breaches of condition, although she expressed the view that the courts would be reluctant to quash the permission:

> E particularly if one takes into account the delay by the authority and our practice with other developers in not requiring compliance before development.

Counsel advised on 17 June 1999 that development commenced in breach of conditions precedent would not have begun pursuant to the outline F permission and detailed consent.

[27] That advice was not disclosed to the appellant. Very remarkably, it was not set out, or even referred to, in the officer's report to the planning committee at their meeting on 18 August 1999, when the pending applications by the appellant came under consideration. The suggested explanation is that Mrs Brown was then away on leave, and the matter G was dealt with by a recent recruit to the legal services department. The report to committee simply referred to the existing permission as valid and went on to note that the development proposed in the outline permission had commenced. The report recommended granting both applications, subject to conditions and a section 106 agreement, and the committee accepted that recommendation.

H [28] All must have seemed to the appellant to be proceeding smoothly, when Aldergate returned to the attack. On 20 October 1999, Aldergate sent a letter before action and, on 5 November 1999, before any planning permission had been granted as a result of the 18 August 1999 decisions, Aldergate and Mr Christian, a local resident living in Camborne Crescent, close to the site, applied, in the High Court, for permission to bring judicial review proceedings to quash those decisions. At the forefront of the grounds set out in their Form 86A application was the contention that the respondents had erroneously assumed that there was already in existence an extant and viable permission. At para 37, it was said:

> Aldergate has very little information as to when, if at all, the conditions of any of the above planning permissions were fulfilled. It will be seen in due course that an (undated) internal memorandum of Bassetlaw suggest[s] that there may well be conditions which were not fulfilled within the required time period. Aldergate's own investigations suggest that Conditions 2, 3, 5 (in part), 7 (in part), 8, 9, 10 and 12 of planning permission 1/95/73 have not been fulfilled within the required time. At the very least this raises questions as to whether any of the planning permissions were extant in August 1999.

[29] Receipt of this Form 86A was acknowledged by the appellant's solicitor on 9 November 1999. On 23 December 1999, Mrs Brown wrote to the solicitor, advising it that, pending certain investigations, no planning permission would be issued pursuant to the August 1999 resolution and that the council could not, at this stage, commit themselves to a position whereby any further works carried out would be assumed to have the benefit of a valid planning permission: "If such works continue it should be clearly understood by your clients that the Council is not at this stage prepared to warrant that those works would be authorised."

The appellant's solicitor demanded to know upon what basis it was contended that the works might not be authorised, to which Mrs Brown replied, on 14 February 2000, saying that:

> A question has been raised over the 1995 permission because of the failure of your clients to comply with the conditions attached to that permission especially conditions precedent and Counsel's advice is being sought on this issue.
>
> Your client's failure to comply with conditions attached to the permission means any further development of the site is a breach of planning control and should your clients continue with the development you should be aware and be on notice that the Council's intention would be to take enforcement action against your clients, which may include an injunction under Section 187B of the 1990 [Act]. By return please provide an assurance that no further development will proceed until the status of the 1995 planning permission in relation to the site has finally been resolved.

In a letter dated 25 February, the appellant's solicitor said:

> We believe that your Council has had all the information required to enable the satisfaction of these conditions. It may be that there was some procedural informality but this reflected the practice of your Council in other matters relating to this development as well as other developments elsewhere. Indeed alterations to the layout have been carried out informally at your Council's request (under condition 2)…
>
> We have now advised our clients to make a new application for approval of all these matters to clear up any future misunderstanding. These applications are made on a without prejudice basis.

[30] A vigorous correspondence followed, but, in any event, on 10 March 2000, applications were made by the appellant in order to seek to meet the terms of conditions 5, 6, 7, 8, 9, 10 and 12 of the detailed approval of 5 December 1995. These were said to be made on a "without prejudice" basis. Subsequently, an application in respect of condition 4 was also made.

[31] Mrs Brown wrote, on 15 March 2000, to the appellant's solicitor, saying that the council intended to seek a declaration in the High Court as to the status of the 1995 planning permission, and her letter continued:

> In the meantime and for the avoidance of doubt Counsel has advised that the development which has already occurred on the site as well as that which appears to be taking place at present is in breach of planning control. Accordingly I must insist that your client desists from developing the site further and provides me with a written assurance that building work on the site has stopped and will not begin again until the Court has given its final judgment as to the status of the planning permission…

[32] The appellant's solicitor responded by pointing out that, as detailed approval of reserved matters had already been granted, its clients could submit applications to meet the outstanding conditions until 1 August 2000. It added that no new development would be undertaken until the recent (10 March) applications were approved.

[33] By now, the respondents had clearly received further advice from counsel, since Mrs Brown referred in a letter dated 27 March 2000 to a number of decisions in the courts, including *FG Whitley & Sons Co Ltd* v *Secretary of State for Wales* (1992) 64 P&CR 296[1], dealing with the effect of a breach of condition upon the starting of development. She observed that it did not seem to be in dispute that development of the site had proceeded in breach of conditions precedent, and questioned whether the council could now properly entertain the recent applications. In reply, the appellant's solicitor pointed out the distinction between approval of reserved matters, which was subject to a three-year time limit, and approvals required under other conditions, which were not. Mrs Brown accepted that and agreed that the council could lawfully deal with recent applications. On the same day, 30 March 2000, the appellant withdrew its 1998 applications for the "compromise" scheme, with the result that Aldergate's application for judicial review was never pursued to a hearing.

[34] There was then further correspondence about the effect of any approval of the March applications and about the need for further information on them, and a meeting took place between officers of the respondents and representatives of the appellant on 7 July 2000. The applications went to the council's planning committee on 19 July 2000, with a recommendation from officers to refuse approval of most of them

[1] [1992] 3 PLR 72

because they were unsatisfactory in a number of respects. The committee approved the applications in respect of conditions 4 and 6, but refused those relating to conditions 5, 7, 8, 9, 10 and 12.

[35] On 31 July 2000, the appellant appealed to the Secretary of State against those refusals. Further details that had been submitted during July in relation to the conditions were considered by the committee on 1 August 2000, but no decision notices were issued by the respondents and no appeals were made by the appellant in respect of these decisions. Also, on 31 July 2000, it submitted two applications, under section 73 of the Act, to vary conditions on the outline permission and the detailed approval. One of these sought to vary condition 1 on the outline permission so as to extend the time for the commencement of development; the other sought to vary the outline permission and detailed approval so that the development of part A of the site could be completed without compliance with conditions 1 to 12. An application was also made under section 73A to obtain retrospective planning permission for the dwellings already built on part of part A of the site. Those applications were said to be made "without prejudice" to the contention that the outline permission would remain valid after 2 August 2000. We were told, during argument, that the first application was in due course refused by the respondents and that there was no appeal by the appellant. The other two applications were not registered by the council because insufficient detail had been submitted to them by the appellant and the appellant took no further action to pursue these applications. It seems that the 63 dwellings that have been built on part A have since been granted full permissions as a result of a number of applications made by the owners or others.

[36] As for the appeal to the Secretary of State against the refusal of approval under the various conditions, that has a somewhat curious history. The appellant's solicitor was told by the Planning Inspectorate that certain missing documents had to be supplied within six months of the LPA's decision. The appellant, however, failed to send the relevant documents by the deadline, and the appeal was, as a result, deemed invalid. Consequently, the conditions in question have never been complied with, save that the details required under conditions 5, 7, and 8 (floor levels, landscaping and hard surfacing) were approved in respect of the developed area within part A of the site.

Legal context

[37] It was accepted by counsel for the appellant in the court below, and not challenged before us, that there is an established principle of law that in order for operations to amount to the commencement of development under a planning permission, those operations must be authorised by the permission in question, read together with its conditions. In general, operations carried out in breach of a condition cannot be relied upon as material operations capable of commencing the development within the meaning of section 56(2). This principle, sometimes referred to as the *Whitley* principle, goes back at least to the decision in *Etheridge* v *Secretary*

A *of State for the Environment* (1983) 48 P&CR 35, where Woolf J referred to the statutory time limit on starting development then contained in para 20(1) of Schedule 24 to the Town and Country Planning Act 1971, and said:

> if it were development that was commenced in contravention of conditions of planning permission, it seems to me that it would be development that,
> B from the planning point of view, would be in breach of a planning control. Therefore it would not be a development contemplated by para 20(1)...

[38] That proposition was expressly approved in *Oakimber Ltd* v *Elmbridge Borough Council* (1991) 62 P&CR 594[1], at p609, by Purchas LJ, with whom Taylor LJ agreed. Beldam LJ was also prepared to endorse it, had it been necessary, in his judgment, to deal with that issue.

C [39] In *Whitley* itself, Woolf LJ, with whom the other two members of the court agreed, reviewed the authorities and formulated the principle at p302:

> As I understand the effect of the authorities to which I am about to refer, it is only necessary to ask the single question; are the operations (in
> D other situations the question would refer to the development) permitted by the planning permission read together with its conditions? The permission is controlled by and subject to the conditions. If the operations contravene the conditions they cannot properly be described as commencing the development authorised by the permission. If they do not comply with the permission they constitute a breach of planning control and for planning purposes will be unauthorised and
> E thus unlawful. This is the principle which has now been clearly established by the authorities. It is a principle which I would have thought made good sense since I cannot conceive that when section 41(1) of the 1971 Act made the planning permission subject to a condition requiring the development to be begun by a specified date, it could have been referring to development other than that which is authorised by the permission.

F

Subsequent authorities have applied this principle. It has also been established that it operates even where the necessary detailed approvals for the works actually carried out have been obtained but where another condition on the permission has to be complied with before development began and there has been no compliance with that other condition: *Leisure*
G *Great Britain plc* v *Isle of Wight County Council* (2000) 80 P&CR 370, where the authorities to this effect are reviewed at pp376-377.

[40] The works carried out by the appellant in the present case were in breach of a number of conditions that had to be satisfied "before development commences". That is common ground. It would seem to follow that, on the face of it, those works could not amount to a valid
H commencement of development under the outline permission. It was indeed the decision of Sullivan J in the court below that the development

---

[1] [1991] 3 PLR 35

authorised by that permission was not commenced prior to 2 August 2000, A
and that the outline permission was no longer extant. He granted
declarations to the respondents to that effect and refused to grant
declarations to contrary effect to the appellant, as well as refusing it a
declaration that it had a legitimate expectation that the respondents would
treat the development authorised by the outline permission and detailed
approval as having commenced before that date, to which the respondents B
were bound to give effect and from which it would be unlawful for the
respondents to resile.

[41] That brings me conveniently to the grounds upon which Sullivan
J's decision is now attacked by the appellant.

Appellant's submissions C

[42] It is contended first by the appellant that, as a result of
the representations made on behalf of the council, and the conduct
generally of the council, it had a legitimate expectation in the terms
that I have just set out. Mr David Holgate QC, for the appellant, submits
that over a period of years from May 1996 the respondents had indicated
that the outline permission had been implemented. He points in particular D
to the meeting during that month, as noted in the letter of 4 July 1996, to
the council's topic paper for the local plan inquiry and to its proof of
evidence used at that inquiry in the autumn of 1998. Reliance is also placed
upon the report to committee in August 1999, with its references to
the development in the outline permission having commenced, as well
as on a number of other documents. All this gave rise, it is said, to E
the expectation on the part of the appellant that the respondents would
treat the development as having validly commenced, despite the breach
of conditions.

[43] Moreover, the respondents gave piecemeal approval to details in
respect of those houses that the appellant in fact built, and took no action
to prevent them from being built. It is submitted that only in their letter of F
14 February 2000 did the respondents give the appellant reason to believe
that they had changed their position on this issue as to the valid
commencement of development under the outline permission. By then, it
was in reality too late for the appellant to rescue its position by putting
together satisfactory applications to meet the outstanding conditions. It
would therefore be unfair and an abuse of power to allow the respondents G
to resile from the position they had adopted for so long a period, especially
since the appellant had relied upon the representation and conduct of the
respondents.

[44] Mr Holgate cites *R* v *Inland Revenue Commissioners, ex parte
Unilever plc* [1996] STC 681 and *R* v *North and East Devon Health
Authority, ex parte Coughlan* [2001] QB 213 as demonstrating how H
a legitimate expression can arise in such circumstances. He recognises
that legitimate expectation may have a more limited role to play in
town and country planning law, because of the extent of public interest
in the planning process. But he submits that the public would have had
very little involvement in how the appellant met the terms of

the conditions in question. As an approval under such a condition is not a "planning permission" under section 57 of the Act, there is no requirement for publicity of or consultation about an application for such an approval: see Articles 8 and 10 of the Town and Country Planning (General Development Procedure) Order 1995. Such an application does not have to be included in the register of planning applications (see Article 25 of the 1995 Order), nor can the Secretary of State direct that such applications be referred to him for decision: section 77 of the Act. The only requirement is that such an application should be in writing: see (*R* v *Flintshire County Council, ex parte Somerfield Stores Ltd* [1998] PLCR 336, at p351, and *Tesco Stores Ltd* v *North Norfolk District Council* [1999] JPL 920.) Therefore, argues Mr Holgate, there is no expectation on the part of the public that there would be consultation or publicity when an application for approval under a condition is made. In the present case, the public might have concerns about the principle of the development of the site, but those could not be raised on such an application because outline permission had already been granted. In any event, the council are entrusted with the responsibility of representing the public interest.

[45] The other way in which this argument is put is that there is a power in the LPA to waive the need for compliance with such conditions, and that, in effect, is what happened in the present case. A developer does not always have to make a formal application under section 73 of the Act to vary or discharge the conditions. The planning system needs to be practical and flexible, as was recognised by Collins J in *Agecrest Ltd* v *Gwynedd County Council* [1998] JPL 325. That case shows that, where an LPA waive the need for such compliance, development can commence, for the purpose of sections 56 and 93 of the Act, without compliance, thus operating as an exception to the *Whitley* principle. There can be exceptions to that principle, as *Whitley* itself established, and as was recognised in *Leisure Great Britain* and in *Flintshire*. Here, the respondents chose, in essence, to deal with the implementation of the permitted development in a phased way, without insisting upon strict compliance with the conditions. Mr Holgate relies upon the Divisional Court decision in *R* v *Secretary of State for the Environment, ex parte Percy Bilton Industrial Properties Ltd* (1975) 31 P&CR 154 as demonstrating that a LPA are entitled to adopt such an approach of phased implementation.

Analysis

[46] It seems to me that all these arguments are interrelated, at least to some extent. If, as Mr Mark Lowe QC submits on behalf of the respondents, we now have a comprehensive code of planning control and the LPA do not have the power to waive, in an informal way, the need for compliance with a condition, the circumstances in which a legitimate expectation of the type suggested here will arise on the part of a developer because of the conduct of the authority will be very exceptional.

[47] It has long been recognised at the highest level that planning law is contained in a comprehensive code. In 1984, in *Pioneer Aggregates (UK) Ltd* v *Secretary of State for the Environment* [1985] AC 132, Lord Scarman, with whom the rest of the House of Lords agreed, emphasised that fact, and said, at p141BC:

> if the statute law covers the situation, it will be an impermissible exercise of the judicial function to go beyond the statutory provision by applying such principles merely because they may appear to achieve a fairer solution to the problem being considered.

[48] The House of Lords was there dealing with the statutory code contained in the Town and Country Planning Act 1971. Since that time, as Sullivan J noted in the present case ([137]), the code has become even more comprehensive. Of particular relevance for present purposes is the introduction, by section 49 of the Housing and Planning Act 1986, of what is now section 73 of the 1990 Act. That provides a statutory mechanism by which a person can, in effect, seek a variation in a condition on a planning permission, or the discharge of such a condition. It does so by providing for an application:

> for planning permission for the development of land without complying with conditions subject to which a previous planning permission was granted. (section 73(1))

[49] Thus, it is the case that (to use the language of *Pioneer Aggregates*) "statute law covers the situation" when a developer wishes to vary or discharge one or more conditions on a permission. As Sullivan J commented, this, together with the power to appeal against a condition in the first place or to obtain retrospective permission under section 73A, builds a considerable degree of flexibility into the statutory code: see [137] to [140] of his judgment. I would add that the reality is that developers of experience, such as the appellant, also generally seek to negotiate the form of conditions with the LPA before any permission is granted.

[50] But the section 73 procedure has an additional importance. Because it operates by means of providing for an application to be made for "planning permission" without certain conditions previously imposed, it imports (as Mr Holgate accepts) the safeguards for third parties and the public generally that apply to applications for planning permission under the Act. Thus, although it is right that an application seeking approval under a condition does not attract the statutory provisions concerning publicity and consultation, an application under section 73 to vary or discharge such an existing condition does. Such an application has to be entered on the planning register, available to the public. Indeed, it can be called in by the Secretary of State for his own determination. The statutory procedure ensures that a condition requiring matters to be dealt with before development begins cannot be varied or discharged without an opportunity for the public to be informed and to object or make other representation. So, the public is engaged in that statutory process.

A [51] In those circumstances, the scope for such variation or discharge to be achieved by some other non-statutory method, bypassing the statutory safeguards for the public, must be extremely limited. Such a change is not simply a matter for bilateral agreement between the developer and the LPA. The public is entitled to be involved. As Dyson J said in the first instance decision in *R* v *Leicester City Council, ex parte Powergen UK plc*
B [1999] 4 PLR 91, at p101F-H:

> section 73 is the provision that parliament has enacted to deal with situations where a developer wishes to develop land without compliance with conditions previously attached to a planning permission. What is required in such circumstances is that the developer apply for planning permission. I do not accept that the provisions of section 73 can be
> C sidestepped by persuading a local planning authority, still less an authorised officer, to vary or waive a condition under the guise of the exercise of a general management discretion in the implementation of planning permissions.

The reasoning of Dyson J was endorsed when that case went to the Court
D of Appeal: see (2001) 81 P&CR 5, at paras 21, 47 and 48. Moreover, it is entirely in harmony with the House of Lords' decision in *R (on the application of Reprotech (Pebsham) Ltd)* v *East Sussex County Council* [2002] UKHL 8; [2002] 4 All ER 58[1], where the issue concerned the process of determining whether development requiring planning permission was involved. At [29], Lord Hoffmann said:

> E It is, I think, clear from this brief summary that a determination is not simply a matter between the applicant and the planning authority in which they are free to agree on whatever procedure they please. It is also a matter that concerns the general public interest and that requires other planning authorities, the Secretary of State, on behalf of the national interest, and the public itself to be able to participate.

F [52] That is very much the nature of town and country planning law. Even more than many areas of public law that concern an individual and a public body, planning law is likely to have to reflect the fact that third parties and the public generally may have interests in any decision. I agree with what was said by Sullivan J, in the present case, at [140]:

> G It is important at all times to remember the public nature of Town and Country Planning. It is not a matter for private agreement between developers and Local Planning Authorities.

[53] The decision in *Agecrest* relied upon by the appellant needs to be seen in its context. The case was dealing with events that had taken place in 1967, before what is now section 73 of the 1990 Act had been enacted. At
H that time, no statutory procedure existed for discharging or varying a condition on a planning permission. The decision can have little relevance

[1] [2002] 2 PLR 60

to cases concerned with events since section 49 of the Housing and Planning Act 1986 came into force. As Richards J said in *Coghurst Wood Leisure Park Ltd* v *Secretary of State for Transport, Local Government and the Regions* [2002] EWHC 1091 (Admin), at [62]:

> I have difficulty in seeing how the decision in *Agecrest* fits into the present statutory framework.

I agree. I would emphasise that, when I referred to *Agecrest* in *Leisure Great Britain*, as an exception to the *Whitley* principle, I was not seeking to endorse it beyond recognising that the decision was there as an authority. For the reasons already indicated, it has no bearing upon the recent appeal.

[54] The scope, therefore, for waiver by non-statutory means of the need to comply with a condition must be extremely limited. That is so whether one is concerned with an alleged waiver of a condition in total or with an allegation that the LPA have allowed development to take place in a phased manner, contrary to a condition. The latter still involves an informal variation of the condition and gives rise to the same problems as any other kind of non-statutory variation. The decision in *Percy Bilton* was, like *Agecrest*, concerned with events before the section 73 mechanism had been introduced by parliament. Moreover, the wording of the condition in *Percy Bilton* was such that it could be construed as meaning that no particular item of development within the permission could be begun until the siting, design, external appearance and means of access of *that* item of development had been approved, rather than that all the details of the whole development had to be approved before any development could begin. On that basis, it is an understandable decision. But however one interprets it, in my judgment it provides no authority today for the non-statutory variation of a condition. The statutory planning system has moved on.

[55] The interests of third parties and the public in such matters also greatly reduce the potential for a legitimate expectation, such as is contended for in the present appeal, to arise. One of the reasons is that it is difficult to see how a legitimate expectation, said to derive from the conduct of the local authority, could operate so as to prevent an interested third party from questioning whether development has validly begun and whether the planning permission is still extant. This is not a remote possibility: the commencement of judicial review proceedings by Aldergate and Mr Christian in the present case demonstrates how such issues extend beyond merely the developer and the local authority. Yet those third parties did nothing to give rise to any legitimate expectation on the appellant's part.

[56] Mr Lowe invited us to say that legitimate expectation could never operate so as to enable the developer to begin development validly and effectively in breach of condition. I am not prepared to adopt so absolute a proposition. It is possible that circumstances might arise where it was clear that there was no third party or public interest in the matter, and a court might take the view that a legitimate expectation could then arise

from the LPA's conduct or representations. But, as was said in *Coghurst Wood*, one suspects that such cases will be very rare. The situation that normally arises in a planning context is very different from that which arises in cases such as *Unilever*, where the issue is essentially one as between the individual and the public body, in that case the Inland Revenue. Legitimate expectation has a far greater role to play in such circumstances.

[57] I turn then to the facts of the present case. It is true that, for a long time, the officers of the LPA assumed that development had started under the outline permission, and this view was expressed upon a number of occasions. Even though they were aware that a number of conditions had not been complied with, they assumed that the works carried out amounted to an implementation of the outline permission. On the other hand, nothing shows that they applied their minds during that period to the legal issue of whether such works could *lawfully* amount to a start of development under that permission. It seems that they, like the appellant's executives and staff, simply assumed that those works had implemented the permission. Neither party appears to have been aware of the *Whitley* principle, until the respondents were advised by counsel on 17 June 1999.

[58] The issue of whether the works carried out in breach of condition amounted to a start to "the development to which the permission relates" within the meaning of section 92(2) of the Act was, and is, essentially a legal one, to be determined, in the last resort, by the courts. It is not simply a matter for the LPA, and it means that any view expressed upon it by the LPA is in a very different category from the normal case of a legitimate expectation that a public body will exercise its powers in a particular way. Moreover, in so far as the doctrine of legitimate expectation is to be seen as "rooted in fairness", as it was put by Bingham LJ in *R* v *Inland Revenue Commissioners, ex parte MFK Underwriting Agents Ltd* [1990] WLR 1545, at p1570, it is relevant that the appellant itself, as a substantial housebuilding company, had access to legal advice, had it wished to take it. It was as capable, as were the LPA, of informing itself as to the legal consequences of commencing development in breach of condition, and of the problems in establishing that this amounted to a start of development under the outline permission.

[59] When these considerations are put into the legal context of the statutory code, I am wholly unpersuaded that any expectation on the appellant's part that its works would be treated as a lawful implementation of the outline permission was a legitimate one. Nor do the respondents' actions amount to an abuse of power. I do not accept that the appellant had no reason to anticipate, until 14 February 2000, that the works carried out on site might not be regarded as being a valid start to the permitted development. It was given a clear warning by the letter of the 23 December 1999 from Mrs Brown, but it should also have been alerted by Aldergate's application for judicial review in early November 1999. That application raised this very point, and while that application was not made by the respondent, the issues that it identified were patently going to have to be given careful consideration by the respondents.

[60] So a warning bell should have sounded some nine months A before the outline permission was due to expire in the event of non-implementation. The evidence does not establish that this was an insufficient period of time for the applicant to have prepared applications capable of securing the approval of the LPA under the outstanding conditions, or, if necessary, of the Secretary of State on appeal. The latter possibility is of relevance because there is an exception to the *Whitley* B principle established by the decision in *Whitley* itself. So long as the developer applies before the deadline for the outstanding approval or approvals, and those approvals are given (either by the authority or by the Secretary of State on appeal) after the deadline, preventing enforcement action from being taken, work done before the deadline in accordance with the scheme ultimately approved can amount to a start to C development. The permission will have been lawfully implemented. The possibility of invoking that exception existed in the present case. The appellant applied in time for the outstanding approvals and, on refusal by the respondents, lodged an appeal with the Secretary of State in respect of one and failed to appeal the other. But for reasons that remain shrouded in mystery, the appellant failed to pursue its appeal, which, in due course D ,lapsed.

[61] This part of the history is relevant to the issue as to whether it would be unfair to allow the respondents to contend that there has not been a lawful start to development. The appellant had a method available by which it might have been able to achieve retrospective authorisation for the works it had carried out on-site. It did not avail itself of that method. E

Conclusion

[62] For the reasons I have set out, the appellant cannot be regarded as having established a legitimate expectation that its works carried out on-site, in breach of condition, would be regarded as a lawful F commencement of development under the outline permission for the purpose of sections 56 and 92(2) of the Act. Nor was there any lawful waiver of conditions, whether generally or to the extent of allowing phased development. Conditions on a planning permission must either be complied with, at least in substance (see *Flintshire*), or, if it is sought to vary or discharge them, the mechanism laid down by parliament in section 73 G of the Act, or in appropriate circumstances in section 73A, must be utilised. That was not done in this case.

[63] In the light of the conclusions that I have reached on the appellant's contentions, it is unnecessary to deal with the argument raised in the respondents' notice that the officers of the LPA have no authority, actual or ostensible, to waive any of the conditions. If the authority had no H power to waive the conditions by non-statutory means, then nor did their officers. That is not to say that the conduct of an authority's officers will be irrelevant to an issue of legitimate expectation. It may be rare that a legitimate expectation of the kind asserted in this appeal will arise in respect of a planning condition, but, if and when it does, the conduct of the

A LPA's officers is likely to be germane to a consideration of the issue. This, however, is not one of those rare cases.

[64] It follows that I would dismiss this appeal.

BODEY J:

[65] I agree.

B **BROOKE LJ:**

[66] I also agree.

Appeal dismissed.

For Product Safety Concerns and Information please contact our EU representative GPSR@taylorandfrancis.com
Taylor & Francis Verlag GmbH, Kaufingerstraße 24, 80331 München, Germany

www.ingramcontent.com/pod-product-compliance
Lightning Source LLC
LaVergne TN
LVHW020637300926
849161LV00004B/739
* 9 7 8 0 7 2 8 2 0 3 7 9 2 *